Computational Microelectronics

Edited by S. Selberherr

Modelling of Interface Carrier Transport for Device Simulation

Dietmar Schroeder

Springer-Verlag Wien GmbH

Dr.-Ing. Dietmar Schroeder
Technische Elektronik
Technische Universität Hamburg-Harburg
Hamburg, Federal Republic of Germany

© 1994 Springer-Verlag Wien
Originally published by Springer-Verlag Wien New York in 1994
Typesetting: Asco Trade Typesetting Ltd., Hong Kong

Printed on acid-free and chlorine free bleached paper

With 69 Figures

ISSN 0179-0307
ISBN 978-3-7091-7368-8   ISBN 978-3-7091-6644-4 (eBook)
DOI 10.1007/978-3-7091-6644-4

Preface

This book represents a comprehensive text devoted to charge transport at semiconductor interfaces and its consideration in device simulation by interface and boundary conditions. It contains a broad review of the physics, modelling and simulation of electron transport at interfaces in semiconductor devices. Particular emphasis is put on the consistent derivation of interface or boundary conditions for semiconductor device simulation. The book is of interest with respect to a wide range of electronic engineering activities, as process design, device design, process characterization, research in microelectronics, or device simulator development. It is also useful for students and lecturers in courses of electronic engineering, and it supplements the library of technically oriented solid-state physicists.

The deepest roots of this book date back to the mid-seventies. Being a student of electrical engineering, who was exposed for the first time to the material of semiconductor device electronics, I was puzzled by noticing that much emphasis was put on a thorough introduction and understanding of the basic semiconductor equations, while the boundary conditions for these equations received very much less attention. Until today on many occasions one could get the impression that boundary conditions are unimportant accessories; they do not stand on their own besides the bulk transport equations, although it is clear that they are of course a necessary complement of these.

The problem of boundary conditions returned to me some years later when our device simulation group at the Technical University of Hamburg-Harburg made some investigations on partitioning the simulation domain with the goal to use transport models of different accuracy (and complexity) in different regions of the device. Here, the problem arose that we had a number of models, materials, and regions, a number of interfaces between the regions, and a number of external boundaries, where we needed consistent interface and boundary conditions for all the occurring combinations. This was the initial event for me to investigate from scratch the origin and nature of interface and boundary conditions. The goal was to obtain a consistent method to systematically derive such conditions for various different situations. In fact, this book is the compiled result of this effort of quite some years.

It was my intention to provide as much detail as necessary for the reader to follow and verify the main reasoning without having to resort to referenced material or having to guess about the subtleties and approximations during a number of un-displayed intermediate mathematical steps. Some basic understanding of semiconductor device physics is required for reading the book, since these topics are only briefly recalled in Chapter 2. Thus, the book should be useful to advanced students, who wish to deepen their understanding of device physics and simulation. Part of the presented matter represents indeed the content of lectures I held in recent years for students of electrical engineering at the Technical University of Hamburg-Harburg. At the same time, the book is intended to support the work of engineers and scientists in the semiconductor industry and research laboratories, who have to deal with problems of interface carrier transport and appreciate a comprehensive collection of the material, with numerous references to continuing literature for probing further.

It is my sincere wish to express my appreciation to the many people who contributed to the book in various ways.

I am extremely grateful to Prof. R. Paul for his encouragement to begin at all the activity of writing this book, for the opportunities he granted to me to carry out the project, and for his careful reading of the manuscript.

My very special thanks are due to my wife Helga for her constant support and encouragement during all these activities. Moreover, she draw mostly all of the figures, and she never got tired to take care of my most pedantic wishes for correction of many details.

I am particularly indebted to S. Selberherr for his interest in this work and for his engagement as the editor of this series.

I like to express my thanks to Prof. G. Lautz for making me enthusiastic for the field of semiconductor electronics by his excellent lectures on electromagnetic fields and solid-state electronics. I would like to thank also Prof. W. Schultz for conveying his style of getting to the roots of a subject, as well as for his inspiring humorous brainwaves.

Many thanks are due to a great number of colleagues all over the world. For many pleasant and interesting discussions I am very indebted to G. Baccarani, F. Odeh, D. Ventura, A. Gnudi, W. Fichtner, G. Wachutka, S. Müller, A. Schenk, J. Nylander, G. Nanz, B. Meinerzhagen, M. Lundstrom, M. Schubert, P. Gough, S. Battersby, R. Kuvecke, W. Schilders, P. Hemker, P. van der Zeeuw, H. Molenaar, W. Joppich, R. Constapel, C. Ringhofer, R. Brunetti, V. Axelrad, J. Bloedel, A. Neureuther, A. Wong, G. Chin, W. Dietrich, A. Hintz, W. Schoenmaker, M. Mosko, K. Jensen, and many others.

I benefited from many fruitful discussions with colleagues and students at the Department of Technical Electronics of the Technical University of Hamburg-Harburg. I particularly appreciate discussions with A. Stelter, P. Conradi, M. Weber, S. Ebmeyer, S. Becker, M. Brandstetter, and O. Kalz. Special thanks are due to my students A. Bergemann, R. Philipps, S. Zschiegner, and T. Ostermann, who carried out much of the numerical work, and helped to clarify many topics by their critical questions.

I like to express very special thanks to Davide, Gipo, Alberto, Antonella, Eleonora, and Zsolt for the warm and friendly atmosphere during a visit of mine for some weeks at the University of Bologna in October 1991.

I close this preface with the hope that the reader will acknowledge and benefit from all the efforts that went into the completion of the book. I have done my best to create this work, and I wish it might be useful to many colleagues in our profession.

Technical University of Hamburg-Harburg   Dietmar Schroeder
Hamburg, August 1993

Contents

Introduction 1

No semiconductor device consists solely of a single material. Important functions are taken over by layers of metal, insulator, or other semiconductors. Successfull device simulation has to account properly for these materials and their mutual interfaces. Besides of the transport equations which the simulator solves in the bulk of the device, physical processes at the interfaces have to be accounted for by corresponding conditions of different materials.

This book concentrates on the models of electron transport at material interfaces, particularly with respect to the simulation of semiconductor devices. It tries to answer questions on what effects at the interface are important for device simulation, how to model these effects properly, and how to include these models into the equations that the device simulator solves. Presently, semiconductor device simulation is applied broadly to the "classical" electronic devices like bipolar transistors or field-effect transistors. Newer device concepts like the "resonant tunneling devices" whose functions depend heavily on quantum effects are still in a research stage. In order to keep the scope of this book sufficiently focused such that the topic can be considered in adequate detail, the book concentrates mainly on these classical devices.

In the context of this book, semiconductor interfaces are considered that are abrupt transitions from a semiconductor to another material. Thus we can identify the possibilities semiconductor-vacuum, semiconductor-insulator, semiconductor-metal, and semiconductor-semiconductor (heterojunction).

Semiconductor interfaces fullfill many important functions in semiconductor devices. Examples are: direction of the flow of electric currents, application of electric potential, isolation of conducting layers, current control in MOSFETs and MESFETs, supply of electric current to the semiconductor (contacts), or rectification of currents (Schottky-diode). Other, less important effects are used to influence the electron transport in the vicinity of the interface. These effects became increasingly important in recent years with the upcoming ability to manufacture high quality semiconductor heterostructures with their possibility to construct many different and complicated structures.

So the intention of the book is to provide a comprehensive review of the physics, modelling and simulation of electron transport at interfaces in semiconductor devices. Particular emphasis is put on the consistent derivation of interface or boundary conditions for semiconductor device simulation. The book combines a review of widely used interface charge transport models with original developments. The main goal is a macroscopic description of the interface in the sense that the effect of the interface on the function of the semiconductor device, i.e. the effect on the transport in the bulk of the device, is of interest. Hence, the situation inside the interface region will only be covered in such detail as is necessary for the development of interface transport models.

After two introductory chapters on the representation of charge transport in the volume and on a general model of the electronic structure of the interface, the book focuses on charge transport at the interface. First, a unified representation of charge transport at semiconductor interfaces is introduced. The following chapters concentrate on one particular interface at a time. These include the semiconductor-insulator interface, the metal-semiconductor contact, and semiconductor heterojunctions. A special chapter is devoted to the interface between gate and channel in a MOSFET because of the particular importance of this structure. In a final chapter, the problem of discretization of the interface equations and their combination with the discretized volume equations is addressed.

Charge Transport in the Volume 2

In this chapter we recall briefly the theory of electron transport in the bulk of the semiconductor. The idea is to give an overview of the hierarchy of the most important transport descriptions in the semiconductor volume. These are the equations to be solved in the material regions that are separated by the interfaces we want to consider. The structure of the volume transport equations and the charge transport mechanisms at the interfaces will then determine the conditions for connecting the volume solutions at the interfaces.

A basic overview of the transport model hierarchy considered is shown in Fig. 2.1. The hierarchy describes electron transport on various levels of detail (and complexity), which emerge from one another by more and more extensive approximations and simplifications. Since the solution of the more complex equations require increased computing resources, one usually chooses that high a level of description as is necessary for a particular problem, while one sticks to the simpler models as long as their level of accuracy is sufficient.

Outgoing from the quantum level of transport descriptions in the top row of Fig. 2.1 we obtain by certain approximations (which are discussed in more detail in Section 2.2 below) the semi-classical description of the Boltzmann equation. Numerous simplified transport models can be derived in turn from the Boltzmann equation, of which the most important are the balance equations derived by the method of moments. The hydrodynamic transport models, which receive increased attention recently, also belong to this class.

After the discussion of the descriptions of transport in the bulk, we will shortly recall the Poisson equation that governs the interaction of the electrostatic potential with the charge distribution. In the final section of the chapter we will discuss the type and number of boundary conditions that are determined by the structure of the volume equations.

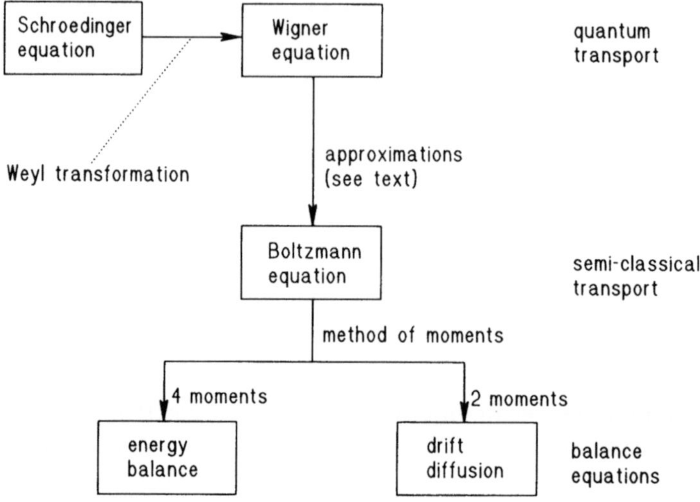

Fig. 2.1 Hierarchy of transport models

2.1 Quantum Transport

Fundamentally, electron transport in crystals has to be described by quantum theory, i.e. the Schrödinger equation. However, for the purpose of semiconductor electron transport in general and the purpose of this book in particular, another (equivalent) representation is preferable. By application of the transformation (Weyl transform)

$$f(\mathbf{x}, \mathbf{k}) = \int d^3 y \, \psi^* \left(\mathbf{x} + \frac{1}{2}\mathbf{y} \right) \psi \left(\mathbf{x} - \frac{1}{2}\mathbf{y} \right) \exp(j\mathbf{k}\mathbf{y}) \tag{2.1}$$

to the wave function ψ of the Schrödinger equation, a function $f(\mathbf{x}, \mathbf{k})$ is defined. Since this function has been introduced for the first time by E. Wigner in 1932 [1], it is usually called "Wigner function". The arguments of the Wigner function are the position vector $\mathbf{x}$ and the wave vector $\mathbf{k}$, which is related to the momentum of the electron. The Wigner function describes the electronic quantum mechanical state, and it is fully equivalent to the wave function. Moreover, it can be easily generalized from pure quantum states to mixed states by substitution of $\psi^*\psi$ in (2.1) with the density matrix ρ [2].

Originally, the Wigner function is normalized as to give unity when integrated over the full phase space $(\mathbf{x}, \mathbf{k})$. This results from the similar normalization property of the absolute square of the wave function ψ. However, for the description of transport processes in semiconductor devices, another normalization is also convenient. In this normalization, the Wigner function is scaled such that its integral over phase space gives the total number of electrons in the considered domain (see also the corresponding discussion of the density matrix normalization in [3]). The integral with respect to **k**-space then gives the electron density (cf. Eq. (2.9) below) instead of the local probability of finding the electron.

For illustrative purposes, I show here only the simple one-particle description as originally published by Wigner [1]. A profound discussion of the Wigner function in the many-particle picture can be found in [4]. A generalization to the case of electrons in crystals, basing the many-particle Wigner function on Bloch and Wannier functions, is given in [5]. In this representation, the total number of electrons naturally comes out of the phase space integral. Also the Pauli principle and electron-electron interactions, which are of course lacking in the one-electron approximation, are inherent in the many-particle formulation.

The advantage of the Wigner function with respect to the wave function is its close analogy with the distribution function in phase space of (semi-) classical transport theory. In particular, macroscopic quantities like the electron concentration n or the electrical current density $\mathbf{J}_n$ are derived from the Wigner function by the same expressions as in the case where f is the classical distribution function (see Section 2.3, Eqs. (2.9)–(2.12)). Another advantage of the Wigner function lies in its ability to simulate semiconductor devices with open boundaries, i.e. boundaries where electrons flow in and out of the device via metallic contacts [5].

The behaviour of the Wigner function is determined by an equation which is derived by applying the Weyl transform to the Schrödinger equation. The resulting quantum transport equation [1]

$$\frac{\partial f}{\partial t} + \frac{\hbar \mathbf{k}}{m} \operatorname{grad} f - \frac{V}{(2\pi)^3} \int d^3k' \, S(\mathbf{x}, \mathbf{k}, \mathbf{k}') f(\mathbf{x}, \mathbf{k}') = 0 \qquad (2.2)$$

will be called "Wigner equation" in the following. It is completely equivalent to the Schrödinger equation. The interaction of the electron with a potential of energy W_{pot} is represented by the third term, where

$$S(\mathbf{x}, \mathbf{k}, \mathbf{k}') = -\frac{2}{V\hbar} \int d^3x' \, W_{\mathrm{pot}}\left(\mathbf{x} - \frac{1}{2}\mathbf{x}'\right) \sin[(\mathbf{k} - \mathbf{k}')\mathbf{x}'] \qquad (2.3)$$

has been used as an abbreviation. The potential energy W_{pot} includes both the long-range macroscopic electrostatic potential as well as the short-range scattering potential caused by the interaction with phonons or impurities.

The Wigner equation (2.2) bears a close similarity to the classical transport equation, the Boltzmann equation, which we shall consider in Section 2.2. Because of this close resemblance, the Wigner function is sometimes also called "quantum distribution function", while the Wigner equation is occasionally termed "quantum transport equation".

2.2 Semi-Classical Transport

In many cases of semiconductor devices, quantum effects are negligible, and a semi-classical description of transport is sufficient. Under the assumptions

- low electron scattering rate;
- small variation of the electrostatic potential in space and time;
- small variation of the electron distribution function with respect to space and wave number,

some approximations can be applied [6] to the Wigner equation, reducing it to the Boltzmann equation [7]

$$\frac{\partial f}{\partial t} + \underbrace{\mathbf{v}(\mathbf{k}) \operatorname{grad} f}_{\text{inertia}} - \underbrace{\frac{q\mathbf{E}}{\hbar} \operatorname{grad}_k f}_{\text{acceleration}}$$

$$+ \frac{V}{(2\pi)^3} \int d^3k' [\underbrace{S(\mathbf{k}, \mathbf{k}')f(\mathbf{x}, \mathbf{k})}_{\text{scatt. out}} - \underbrace{S(\mathbf{k}', \mathbf{k})f(\mathbf{x}, \mathbf{k}')}_{\text{scatt. in}}]$$

$$= 0. \tag{2.4}$$

Thus, the Boltzmann equation is the semi-classical limit of the Wigner equation, and the Wigner function in this limit becomes the distribution function of semi-classical transport theory.

Note that it is not necessary to invoke the concept of wave packets for deriving the Boltzmann equation from quantum transport theory, as is sometimes stated in the literature. The concept of wave packets does not make much sense if the distribution function describes a large number of distributed electrons, as is usual in semiconductor devices.

The quantity $\mathbf{v}(\mathbf{k})$ in (2.4) is the group velocity, which is

$$\mathbf{v} = \frac{\hbar \mathbf{k}}{m} \tag{2.5}$$

for spherical parabolic bands, where m is the effective mass. Eq. (2.4) is also used commonly for non-parabolic bands by replacing (2.5) by the respective group velocity [8]

$$\mathbf{v} = \frac{1}{\hbar} \operatorname{grad}_k W(\mathbf{k}), \tag{2.6}$$

where $W(\mathbf{k})$ is the energy dispersion of the band.

The force on an electron in the third term of (2.4) is expressed here as the effect of the electric field $\mathbf{E}$ on a negative charge. This is valid only if the

semiconductor region is homogeneous; otherwise, additional forces occur that are related to changes in the band structure [9], expressed e.g. by gradients of the electron affinity or the effective mass [10, 11].
The scattering rates in (2.4) include a factor $1 - f$, corresponding to the target state of the scattering, which takes care of the Pauli principle in the case of degeneration by making the scattering probability vanish if the target state is occupied.
The above statements on the normalization of the Wigner function remain valid for the distribution function, too. The distribution function $f(\mathbf{x}, \mathbf{k})$ gives the probability of finding an electron with wavevector $\mathbf{k}$ at location $\mathbf{x}$. The qualitative difference between the Wigner function and the distribution function is that, while the value of both functions is a real number, the distribution function is always positive and can thus be interpreted as a probability. The Wigner function, however, can also take on negative values, and the possible strong oscillations between positive and negative values are responsible for the quantum effects of interference and cancellation.
In most applications of device simulation, the interaction of the charge carriers with the magnetic field is not important. Hence, we disregarded the magnetic field in the above introduction of the Boltzmann equation. If the magnetic field has to be considered, the Lorentz force appears as an additional driving term in (2.4) [12].
A physical interpretation of the Boltzmann equation can be made as follows (cf. Fig. 2.2). As indicated in Fig. 2.2, the temporal change of f depends on the balance of fluxes into and out of a volume element in $\mathbf{x} - \mathbf{k}$ space (phase space) at position $\mathbf{x}$ and wave vector $\mathbf{k}$. The second term of (2.4) describes the inertial motion of a particle, corresponding to a flux parallel to the x-axis in Fig. 2.2. The third term describes acceleration by a

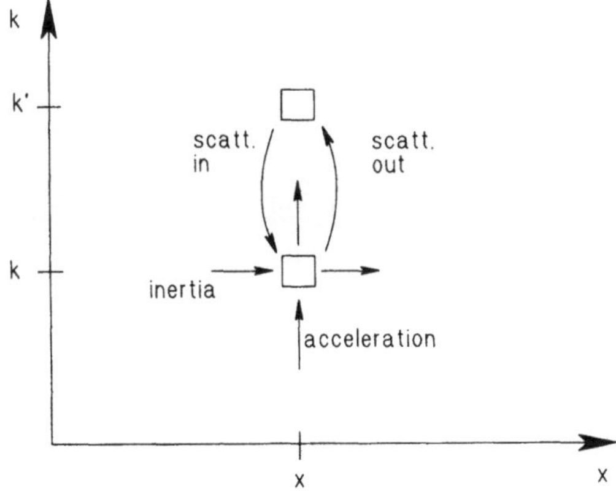

Fig. 2.2 Flux balance of a volume element in phase space

gradient of the potential energy, which corresponds to a movement parallel to the k-axis in Fig. 2.2. The fourth term describes scattering out of the $(\mathbf{x}, \mathbf{k})$ phase space element to another one at $(\mathbf{x}, \mathbf{k}')$, while the last term describes the scattering from $\mathbf{k}'$ into $\mathbf{k}$. These processes are also shown in Fig. 2.2.

In summary, the Boltzmann equation states the balance of particle fluxes in an element of phase space. In the discussion of transport at semiconductor interfaces, we will return to this picture in order to obtain the flux balance at the interface.

If several bands, labeled by the band index v, are considered, a distribution function f_v for each band is introduced. These distribution functions are determined by a Boltzmann equation for each band, which correspondingly reads

$$\frac{\partial f_v}{\partial t} + \mathbf{v}_v(\mathbf{k}) \operatorname{grad} f_v - \frac{q\mathbf{E}}{\hbar} \operatorname{grad}_k f_v$$

$$+ \frac{V}{(2\pi)^3} \sum_{v'} \int d^3k' [S_{vv'}(\mathbf{k}, \mathbf{k}')f_v(\mathbf{x}, \mathbf{k}) - S_{v'v}(\mathbf{k}', \mathbf{k})f_{v'}(\mathbf{x}, \mathbf{k}')]$$

$$= 0. \tag{2.7}$$

Here, the group velocity $\mathbf{v}_v(\mathbf{k})$ of the respective band has been used. These equations include scattering between different bands by additional scattering terms $S_{vv'}$, thus introducing generation-recombination or intervalley scattering into the transport description.

The distribution function f_v we considered so far describes the state of the *electrons*, regardless of the nature of the band v. In case of a valence band ($v = p$), i.e. if the electron effective mass of the band is negative [8], it is convenient to rewrite the Boltzmann equation in terms of a fully equivalent *hole distribution function*. Since f is the probability of finding an electron at the respective point in phase space, $1 - f$ is the probability of *not* finding an electron, i.e. finding a hole. Hence we introduce the hole distribution function g by writing

$$g_p(\mathbf{k}) = 1 - f_p(-\mathbf{k}). \tag{2.8}$$

The minus sign in the argument of f_p reflects the fact that holes travel in the opposite direction as the electrons with the same wave vector. Inserting g into the Boltzmann equation yields an equation with an identical structure; however its interpretation is now in terms of holes as an independent kind of particle instead of electrons.

In principle, the electron energy is a function of wave vector *and* position, i.e. like the distribution function a function defined in the six-dimensional phase space. In the semiconductor volume, this function changes only smoothly with position, as the electrostatic potential distribution shifts the band structure. Hence, in the *bulk* of a semiconductor device, the aforementioned assumptions for deriving the Boltzmann equation are sufficiently satisfied in most cases, and one is allowed to use the Boltzmann equation for a description of electron transport. In the vicinity of an *interface*, however, this is no longer true, and one has to be careful about the validity of the models used. Luckily, quantum effects in the neighbourhood of interfaces are often localized to a thin boundary layer near the interface. Due to its thinness, the boundary layer can be approximated by a surface with zero thickness, and the transport effects in the layer can be included in the boundary condition. Thus, the semi-classical bulk transport model can be used almost everywhere in the device, and the quantum effects near the interface are incorporated into a model of the interface. The elaboration of this principle will be the guiding line for the whole book.

2.3 Balance Equations

In most cases of practical usage of device simulation, e.g. process and device development, the solution of the Boltzmann equation is by far too costly in terms of computer power and turnaround time. If necessary, the full solution of the Boltzmann equation is done by the statistical Monte-Carlo method (for a review see [7]) where the history of many thousands of electrons is simulated in order to obtain a sample of the full electron system. On the other hand, in many cases a solution on such a level of detail is not necessary, and approximative transport models give fully sufficient results. Thus, for engineering purposes of device simulation, approximations of the Boltzmann equation are solved which describe the transport physics at an appropriate level of accuracy.

One of the main problems with a numerical solution of the Boltzmann equation is the dimension of the space of independent variables, which can amount to up to $3 + 3 + 1$. Without regard to time, a discretization into 100 elements on each axis of phase space (which is not all too much) results in 10^{12} discrete unknowns, which is far beyond the capabilities of today's computers. Hence, approximation methods for the Boltzmann equation usually aim at the reduction of the dimension. Since the geometrical aspect of the problem under consideration is normally fixed, there is not much room for reduction in the positional x-subspace for a given device. Thus, the approximation efforts usually concentrate on simplifications with regard to the k-subspace [13]. In general, this is done by introduction of certain assumptions or expansions with respect to the k-variable of the distribution function. The computation of the expansion coefficients then involves integration with respect to the total k-space, or subspaces of it.

By application of an approximation method of this kind to the Boltzmann equation, one arrives at a set of approximate transport equations (cf. Fig. 2.1). The flux balancing character of the Boltzmann equation is preserved through the approximation procedure to some extent, and results in the conservation of certain averaged quantities (e.g. particle number or energy). Hence, the transport equations can usually be interpreted as a set of balance equations.

The principle of the method is described in [14], Chapter VII. A typical procedure for deriving an approximation of the Boltzmann equation is the following "generalized" method of moments:

1. Assume a functional form for $f(\mathbf{x}, \mathbf{k})$ with position dependent parameters;
2. Insert it into the Boltzmann equation;
3. Multiply with $M(\mathbf{k})$, where $M(\mathbf{k})$ determines the weight function (e.g. 1, k_x, k^2);
4. Integrate with respect to $\mathbf{k}$ (i.e. take the moment).

The result of this procedure is a system of partial differential equations for the position dependent parameters. If the number of projections (moment integrations) equals the number of unknowns (the parameters), this equation system constitutes a self-consistent approximate transport model derived from the Boltzmann equation. In like manner, the equations are sometimes called "moment equations" or "balance equations".

The various transport models differ from each other by the functional form assumed for the distribution function (shifted Maxwellian [15], polynomial asymmetric part [16], spherical harmonic functions [17]), by the number of parameters (concentration, current density, electron temperature), the number and form of the weight function (powers of k, spherical harmonics), and the part of $\mathbf{k}$-space to integrate over (full space, solid angle with fixed $|\mathbf{k}|$).

There exist other approximation methods which do not explicitly make use of assumptions of the form of the distribution function, but rather evaluate the result of the moment integration of the Boltzmann equation in terms of other moments. This approach does not lead directly to a self-consistent set of equations (i.e. an identical number of equations and unknowns), but requires additional assumptions for "closing" the equation system (closing relations), or when evaluating the scattering integrals. In some sense, these additional assumptions are equivalent to an implicit presumption of a functional form of the distribution function. Thus, the two approximation methods may be regarded as two incarnations of the same principle.

As mentioned in Section 2.1, macroscopic quantities are deduced from the distribution function by evaluating moments, i.e. averages with respect to $\mathbf{k}$. The first two moments (note that the *first* moment is of order *zero* etc.) of the distribution function are

$$n(\mathbf{x}) = \frac{1}{4\pi^3} \int_{-\infty}^{\infty} d^3k \, f(\mathbf{x}, \mathbf{k}), \tag{2.9}$$

$$\mathbf{J}_n(\mathbf{x}) = -q \frac{1}{4\pi^3} \int_{-\infty}^{\infty} d^3k \, \mathbf{v}(\mathbf{k})f(\mathbf{x}, \mathbf{k}), \tag{2.10}$$

while the 3rd and 4th moments are

$$w(\mathbf{x}) = \frac{1}{4\pi^3} \int_{-\infty}^{\infty} d^3k \, W(\mathbf{k})f(\mathbf{x}, \mathbf{k}), \tag{2.11}$$

$$\mathbf{Q}(\mathbf{x}) = \frac{1}{4\pi^3} \int_{-\infty}^{\infty} d^3k \, \mathbf{v}(\mathbf{k})W(\mathbf{k})f(\mathbf{x}, \mathbf{k}). \tag{2.12}$$

These expressions constitute the relationships between the distribution function and the electron concentration n, the current density $\mathbf{J}_n$, the energy density w, and the energy current density $\mathbf{Q}$. They have the form of averages with respect to $\mathbf{k}$. In case of a parabolic band structure, where the energy is proportional to k^2 and group velocity is proportional to $\mathbf{k}$, the expressions take on the form of moments in the original sense, i.e. averages of powers of k. Here, we use the term "moment" also in the more general sense of Eqs. (2.9)–(2.12).

It is important to note that the even moments of the distribution function give conserved quantities, while the odd moments yield the corresponding fluxes of those.

The equations derived by the method of moments as described above are the balance equations of the conserved quantities. As a result of the approximation process, the balance equations are still defined only on the (at most) three-dimensional position space. The original smooth $\mathbf{k}$-dependence of the distribution function has transformed into a set of discrete quantities. In this way the reduction of complexity, as desired at the beginning of this section, is accomplished.

In order to enhance the clarity of the above explanations, an example of the method of moments is demonstrated in the following [18]. For the distribution function, we choose the form

$$f(\mathbf{x}, \mathbf{k}) = \exp\left(-\frac{1}{k_B T}\left[\frac{\hbar^2 k^2}{2m_n} + W_c(\mathbf{x}) - \phi_n(\mathbf{x})\right]\right)$$
$$\times \left[1 - \frac{\tau}{k_B T}\frac{\hbar}{m}\mathbf{k} \cdot \mathbf{a}(\mathbf{x})\right], \tag{2.13}$$

which is based on a perturbative solution of the Boltzmann equation. Basically, it consists of the equilibrium Boltzmann distribution augmented by an additional term responsible for current flow in non-equilibrium. The free parameters of (2.13) are the quasi-Fermi energy $\phi_n(\mathbf{x})$ and the vector $\mathbf{a}(\mathbf{x})$. They are related to electron concentration n and current density $\mathbf{J}_n$ by the moments (2.9)–(2.10). The electron concentration is obtained from the distribution function by taking the zero order moment, i.e. inserting (2.13) into (2.9). The evaluation yields

$$n = N_c \exp\left(\frac{\phi_n - W_c}{k_B T}\right), \tag{2.14}$$

where N_c is the effective density of states [19]

$$N_c = \frac{1}{4\pi^3}\left(\frac{2\pi m_n k_B T}{\hbar^2}\right)^{3/2}. \tag{2.15}$$

Insertion of (2.13) into (2.10) gives the current density $\mathbf{J}_n$ in terms of $\mathbf{a}$:

$$\mathbf{J}_n = \frac{q\tau}{m_n} N_c \mathbf{a} \exp\left(\frac{\phi_n - W_c}{k_B T}\right). \tag{2.16}$$

So we have introduced two unknowns in the assumption (2.13), one scalar variable ϕ_n and one vector variable $\mathbf{J}_n$ (or, equivalently, n and $\mathbf{a}$). Consequently, we need one scalar equation and one vector equation for these variables. These are obtained by the method of moments. Taking the zero order moment of the Boltzmann equation, after inserting (2.13) into (2.4), results in the equation

$$\frac{\partial n}{\partial t} - \frac{1}{q}\,\mathrm{div}\,\mathbf{J}_n = 0. \tag{2.17}$$

This equation expresses particle conservation; it is the well-known continuity equation of semiconductor device theory (in the absence of generation or recombination). The continuity equation (2.17) determines the electron concentration n or, likewise, the quasi-Fermi energy ϕ_n.
The first order moment of the Boltzmann equation gives

$$\frac{\partial \mathbf{J}_n}{\partial t} - \frac{qn}{m_n}\,\mathrm{grad}\,\phi_n + \frac{\mathbf{J}_n}{\tau_p} = 0. \tag{2.18}$$

Correspondingly, (2.18) as a vector equation determines $\mathbf{a}$ or the current density $\mathbf{J}_n$, and expresses the balance of momentum. The quantity τ_p is the momentum relaxation time; it results from the procedure as an integral involving the scattering terms of the Boltzmann equation (details can be found in [18]).
In many cases, the order of magnitude of τ_p is relatively small compared to the time scale of interest. An additional quasi-static approximation of (2.18) can be made in this case, resulting in

$$\mathbf{J}_n = \mu_n n\,\mathrm{grad}\,\phi_n, \tag{2.19}$$

where the mobility μ_n results as

$$\mu_n = \frac{q\tau_p}{m_n}.$$ (2.20)

Eqs. (2.17) and (2.19) constitute the classical drift-diffusion transport model. Thus, the original complexity of (1 unknown, 6 dimensions) of the Boltzmann equation has been reduced to (4, 3) in this example.
If diffusion currents are negligible, the band edge W_c and the quasi-Fermi level are parallel, and the current density becomes

$$\mathbf{J}_n = \mu_n n \operatorname{grad} W_c = q\mu_n n \mathbf{E} \quad \text{(drift)}.$$ (2.21)

In the case of submicron devices, where hot-electron effects have to be taken into account, additional parameters can be introduced into (2.13), e.g. by allowing the electron temperature T to vary with position. (A respective distribution function will be considered in Chapter 7, cf. Eq. (7.7).) Additional equations for these parameters are obtained analogously by taking the respective higher moments of the Boltzmann equation, resulting for instance in the energy balance equation. Thus, a (hopefully) closer approximation of the true distribution function is obtained by increasing the discrete number of unknowns in the equation set of the transport model.
The assumption (2.13) is valid only in the non-degenerate case. If the electrons are in degeneration, the distribution function of (2.13) has to be modified, introducing the Fermi-Dirac distribution. The exact formulation can be found in e.g. [18, 20]. In this case, the zero order moment of the distribution function can be written as [21]

$$n = \gamma_n N_c \exp\left(\frac{\phi_n - W_c}{k_B T}\right),$$ (2.22)

where the "degeneration factor" γ_n is defined by

$$\gamma_n = F_{1/2}\left(\frac{\phi_n - W_c}{k_B T}\right) \exp\left(-\frac{\phi_n - W_c}{k_B T}\right).$$ (2.23)

$F_{1/2}$ is the Fermi integral of order $\frac{1}{2}$ in the Dingle convention [22]. Note that γ_n goes to unity in the case of non-degeneration. Hence, (2.22) represents a respective generalization of (2.14). The equivalent relationships for holes are [21]

$$p = \gamma_p N_v \exp\left(\frac{W_v - \phi_p}{k_B T}\right)$$ (2.24)

$$\gamma_p = F_{1/2}\left(\frac{W_v - \phi_p}{k_B T}\right) \exp\left(-\frac{W_v - \phi_p}{k_B T}\right),$$ (2.25)

where N_v is the effective density of states in the valence band [19].

2.4 Poisson Equation

The Boltzmann equation in Section 2.2—and thereby also the transport models derived from it—describe the reactions of charge carriers to the driving force of the electric field. Hence, the transport equations constitute in principle equations for those unknowns that give information on the behaviour of the carriers (e.g. the distribution function for the Boltzmann equation, or n and $\mathbf{J}_n$ for the drift-diffusion equations). In order to obtain a self-consistent formulation, the transport equations have to be complemented by the equation that determines the electric field.

This is given by the third Maxwell equation, which relates the electric field to the electric charges [23]. It reads

$$\operatorname{div} \mathbf{D} = \rho, \tag{2.26}$$

where $\mathbf{D}$ is the electric displacement and ρ the charge density. In essence, (2.26) tells that the electric charge is the source of the electric displacement. In a linear dielectric medium, the electric field $\mathbf{E}$ is related to $\mathbf{D}$ by [23]

$$\mathbf{D} = \varepsilon \mathbf{E}, \tag{2.27}$$

with the dielectric constant ε of the medium. The electric field $\mathbf{E}$ in turn is related to the electrostatic potential φ by [23]

$$\mathbf{E} = -\operatorname{grad} \varphi. \tag{2.28}$$

Finally, by inserting (2.27) and (2.28) into (2.26), we arrive at the Poisson equation

$$\boxed{\Delta \varphi = -\frac{\rho}{\varepsilon}.} \tag{2.29}$$

Since the potential φ is usually one of the unknowns to be solved for in device simulation, (2.29) is the most convenient formulation of the relationship between φ and the electric space charge for the present purpose.

In Section 2.5 it will become clear that also for the Poisson equation the introduction of connecting conditions at a semiconductor interface is necessary. The derivation of these interface conditions will be the main topic of Chapter 3.

2.5 Necessity of Boundary Conditions

It can be seen from the preceding sections that all the transport models presented, ranging from the Wigner equation over the Boltzmann equation to the moment equations, are containing differential operators. It is well known that equations containing differential operators only have unique solutions if they are augmented by conditions on the boundaries [24], where the order of the differential operator corresponds to the number of required boundary conditions. This requirement can be made plausible easily. The existence of a differential operator means that the solution of the equation at a particular point depends on the solution at the neighbouring points. If the particular point lies on the boundary of the domain, a neighbouring point does not exist on the boundary side, and the differential equation has to be substituted in that respect by the boundary condition. Thus, with respect to the question of boundary conditions, we have to consider the regions where the equations and solutions are defined upon.

We start with a basic region—let us call it "world". The world is a closed system. This means that no interaction with the outside exists, and all things of interest are inside. From the closure of the world follow some properties of the possible boundary conditions. In case of an infinite world, the condition usually is that no physical effects occur at infinity (e.g. in electrostatics the electric field must tend to zero when going to infinity). In case of a finite world, no transport of anything must occur across the boundary (cf. Fig. 2.3). Sometimes, a cyclic world is considered where the solution on one end determines the solution on the other end (e.g. the Born-von Karman boundary conditions in solid-state physics [25]).

The closed system as a whole can only be treated in simple cases where just a few things are present in an otherwise empty space. In the case of the simulation of electron devices, which are part of a complex circuit, it is in general not possible to consider the system as a whole. Instead, only a small section of the total world, namely the electron device, is of interest. Thus, a partitioning of the world into a "region" of interest and the "environment", as sketched in Fig. 2.3, appears to be convenient.

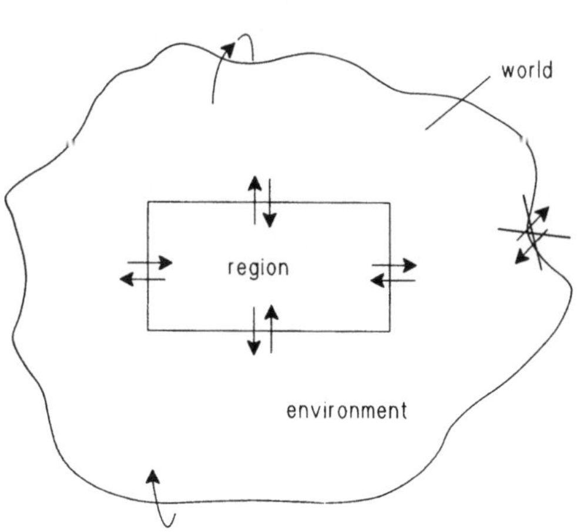

Fig. 2.3 Partitioning of the "world" into a region and an environment

This partitioning has certain consequences. Normally, the region is not a closed system, but interacts in some way with the environment. Now we are in the following dilemma: Because of the complexity of the whole world, we want to consider only a region; but the region interacts with the rest of the world, which we do not want to consider because of its complexity. The solution is that, from the viewpoint of the region, the environment is replaced by a *boundary condition* for the region. In this sense, the boundary condition can be seen as a simple model of the environment and the interaction mechanisms with the environment. The environment is thus reduced to a model of its interaction with the region at the boundary. (In order to be able to describe the interaction with the rest of the world by a boundary condition, it is necessary that the interaction is local at the boundary. This is usually the case in semiconductor devices.)

Sometimes it is convenient to further divide the region into "subregions". This is the case if the region consists of different materials, or if the regions are characterized by different models. Also the distribution of the simulation domain on the numerous processors of a parallel computer requires a decomposition into subregions. The partitioning leads to the introduction of interfaces that separate the subregions, as sketched in Fig. 2.4. It is also indicated in the figure that the subregions interact with each other at the interface. These interactions, in turn, have to be described by respective *interface conditions*.

According to this discussion, we have to distinguish *boundaries* (external boundaries) from *interfaces* (internal boundaries). Internal boundary means that the domain of interest is intersected by the interface, i.e. the distribution function is of interest on both sides of the interface. This may be the case, for instance, at the heterojunction in a semiconductor heterojunction device. An external boundary, on the other hand, represents the edge of the domain of interest to the outer world, i.e. the distribution function is of interest only on the interior side of the boundary. For instance, the metal contact to a semiconductor device is usually considered as an external boundary.

It should be made clear that a subtle distinction between boundaries and

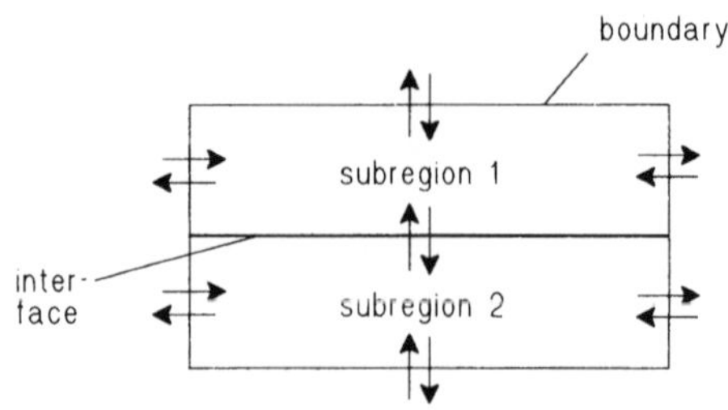

Fig. 2.4 Subregions, interfaces, and boundaries

interfaces exist. In case of a boundary, the environment behind the boundary is not under consideration; instead a simple model of the environment is used as a boundary condition. In case of an interface, however, a solution on both sides of the interface is sought. While in the boundary condition only the influence of the external world on the processes in the region are described, in case of an interface the influences of both subregions upon each other have to be expressed. Thus we state that the *boundary condition* describes a one-way effect (the effect of the environment on the region), while the *interface condition* describes a mutual effect between the subregions. This distinction implies that in the case of an internal boundary two conditions are necessary (one for each side), while in the case of an external boundary only one condition for the interior side must be specified. Hence we find that for identical transport models we need twice the number of equations on an interface than on a boundary.

Usually, the *mathematical* interfaces considered in this section coincide with the types of *physical* interfaces as defined in Chapter 1. For instance, in the simulation of a semiconductor device, only the situation inside the semiconductor region may be of interest. Hence, the material interfaces between the semiconductor and the surrounding metal or insulator would be chosen as the boundaries of the simulation region. The material interfaces between regions of different kinds of semiconductor (heterojunctions) have to be treated as mathematical interfaces because of the abrupt change of parameters. In order to set up boundary conditions as well as interface conditions in the above defined sense, the carrier transport at the respective material interfaces needs to be investigated. This will be the main task of the chapters to follow.

2.5.1 Enumeration of Boundary Conditions for the Boltzmann Equation

Considering the Boltzmann equation, we note that (2.4) is of first order in $\mathbf{x}$ and $\mathbf{k}$ each. The same is true for the Wigner equation (2.2). Hence *one* boundary condition with respect to $\mathbf{x}$ and *one* boundary condition with respect to $\mathbf{k}$ is necessary. Mathematically, the structure of the Boltzmann equation demands that boundary conditions have to be specified on those parts of the boundary where the characteristics point *into* the domain [26]. Let us consider the boundaries in position space first (cf. Fig. 2.5). If the region is finite, *two* boundaries exist for each dimension. Hence, the specification of a condition on the whole boundary would mean that we prescribe a total of *two* conditions. Since only *one* condition is necessary, we would overdetermine the problem. The correct way to deal with this situation is the specification of boundary conditions only on those portions of the boundary where an *inflow* of particles occurs [13]. The inflow boundaries

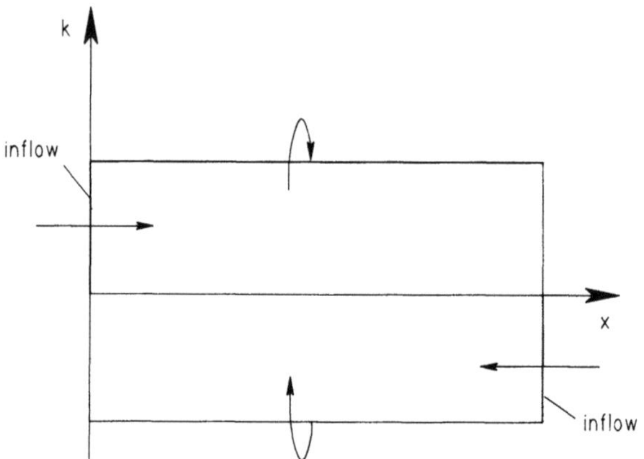

Fig. 2.5 Boundaries in phase space

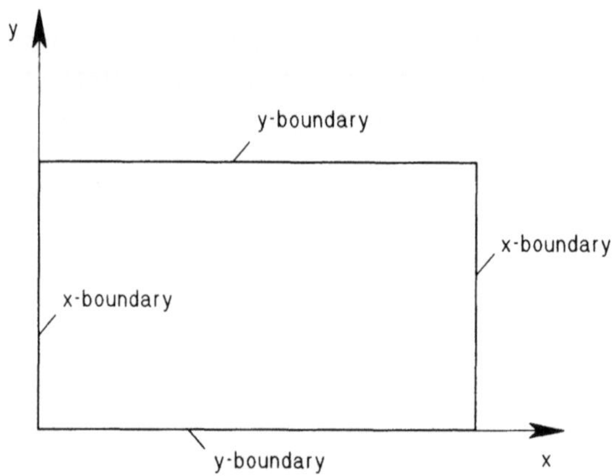

Fig. 2.6 Boundaries in position space

in **x**-subspace are those where the normal component of the particle velocity points inwards [26]. This rule is consistent with the aforementioned condition concerning the characteristics.

The inflow regions on the x-boundary are indicated in Fig. 2.5. It is apparent from the figure that in a sense we have half a boundary condition on two boundaries each; this adds up to one boundary in total, as required.

In **k**-subspace, the domain of interest is bounded by the surface of the first Brillouin zone. The usual boundary condition is periodicity of the distribution function on the zone edges (with the repeated zone scheme [25] in mind). In the picture of the reduced zone scheme [25], this condition turns into a cyclic boundary condition (cf. Fig. 2.5). The condition satisfies the characteristics rule, because it can be interpreted as specifying the inflow on one surface of the Brillouin zone by the outflow on the corresponding opposite surface. Thus, we have one boundary condition for each k-dimension, as required.

In many cases, the distribution function goes exponentially to zero when approaching the boundary of the Brillouin zone. Thus, often it is usual practice to disregard the zone edge at all, and to consider an infinite k-space instead. The only requirement then is the asymptotic approach of the distribution function towards zero; if this condition is satisfied no boundary conditions need to be considered further.

2.5.2 Enumeration of Boundary Conditions for the Poisson and Balance Equations

Since the Poisson and the balance equations do not contain k-dependencies anymore, we need only to consider boundaries in position space.

The Poisson equation (2.29) contains a second order differential operator with respect to position. Hence *two* boundary conditions with respect to each x-direction are necessary. Consider Fig. 2.6, where a two-dimensional domain in position space is shown. If we specify one condition on each boundary, the requirement of two boundary conditions per dimension is exactly satisfied. According to Fig. 2.6, this is equivalent to the specification of one boundary condition on the total boundary of the domain.

Turning now to the balance equations introduced in Section 2.3, we notice that the current equation (2.19) as well as the continuity equation (2.17) contain a first order differential operator. This adds up to two differential equations of first order, requiring a total of two boundary conditions for each space dimension. Alternatively, we can insert the current equation into the continuity equation (hereby eliminating the current density), ending up with a single differential equation of second order. This formulation needs two boundary conditions for each dimension, too. Hence we have a similar situation as with the Poisson equation. We find that for *each pair of transport equations*—i.e. one conservation and one current equation—the specification of one boundary condition on the total boundary of the domain is required. This is also true for pairs of transport equations derived from higher moments of the Boltzmann equation (e.g. the energy conservation equation and the energy current density).

3 General Electronic Model of the Interface

In this chapter we introduce a general basic understanding of the situation that electrons, which are moving through a crystal, find when they encounter a material interface. To stress an analogy from the world of theatre, in this chapter we prepare the stage and set up the scenery for the play called "electron transport", which will be performed in the following chapters. The concept developed in this chapter will allow some statements to be made on the charge distribution at the interface. The influence of these charges on the electrostatic potential then will lead to the basic interface conditions for the Poisson equation.

We shall content ourselves with a simple, phenomenological model conception that suffices for setting up the models required by device simulation. We will keep mainly with a more macroscopic model, touching the microscopic phenomena only briefly. A comprehensive overview of the microscopic treatment of interfaces and surfaces can be found for instance in the book by Bechstedt and Enderlein [27].

3.1 Energy Band Structures at the Interface

In Chapter 2 we based the description of electron transport in the volume on the Boltzmann transport equation. In order to investigate the effect of an abrupt material interface, we have to pay attention to the question what ingredients of the Boltzmann equation change when we cross the interface. Thus, we have to identify the essential properties of the Boltzmann equation that depend on the material species. Looking at (2.4), we find that these are on one hand the scattering mechanisms, and on the other hand the energy band structure $W(\mathbf{k})$, which is implicitly contained in the group velocity $\mathbf{v}(\mathbf{k})$ (see Eq. (2.6)) and in the acceleration force.

In (2.4), it is assumed that the distribution function and the potential energy vary only smoothly in the position space. Otherwise it would not be possible to use spatial gradients of these quantities when setting up the flux balance depicted in Fig. 2.2. This assumption is violated if the phase space element is adjacent to a material interface. While the scattering terms for such a phase space element remain the same, the flux coming from that side where the interface lies cannot be determined by a gradient, but must be supplied by a respective interface condition. This behaviour will be considered in detail in Chapter 4.

Thus we conclude that the scattering terms stay mainly unaffected by the presence of the interface, while the inertia and acceleration terms of (2.4) require special attention because of the change of the band structure. If we go from the first material across the interface to the second material, we find the energy band structure of the first material at the start and the band structure of the second material at the end of this path. At the interface, a transition between the two energy band structures must have taken place. In first order of approximation, we can assume that this transition occurs abruptly at the interface. As a result, we state that primarily *the model of an interface consists of an abrupt change of the energy band structure.*

Before we proceed with taking a look at some examples, we have to define the notation for the band structure that will be used subsequently. The full band structure of a material i is described by a set of functions $W_{v,i}(\mathbf{k}_i)$, which relate the energy of an electron in the crystal to the wave vector. The index v denotes the individual bands of the band structure, as e.g. $v = c$ for the lowest conduction band or $v = v$ for the highest valence band. In order to fix the origin of the energy scale for the band structure of a particular species, we (somewhat arbitrarily) define the minimum of the lowest conduction band as zero energy. However, this choice does not say anything about the actual energy of an electron at some location in a semiconductor device; it only serves to fix the *relative* position of the bands of different materials on the energy scale.

In principle, the electron energy is a function of wave vector *and* position, i.e. a function defined in the six-dimensional phase space. In the semiconductor volume, this function changes only smoothly with position, as the electrostatic potential distribution shifts the total band structure continuously. At the interface, the band structures of the two materials are face to face with each other; we encounter an abrupt change of the band structure. As we will see later, the shift of the band structure of one material with respect to the other plays an important role for the electron transport at the interface. By the above choice of the energy origin for each species, the offset between the total band systems of each of the two materials equals the conduction band difference ΔW_c.

As a first example, we consider the semiconductor-vacuum interface. Figure 3.1 shows the energy dispersion relationships of the two regions that face each other. We see the valence and conduction bands of the semiconductor and the single ("conduction") band of the vacuum. The different curvatures of the bands indicate different effective masses of the bands. The band offset

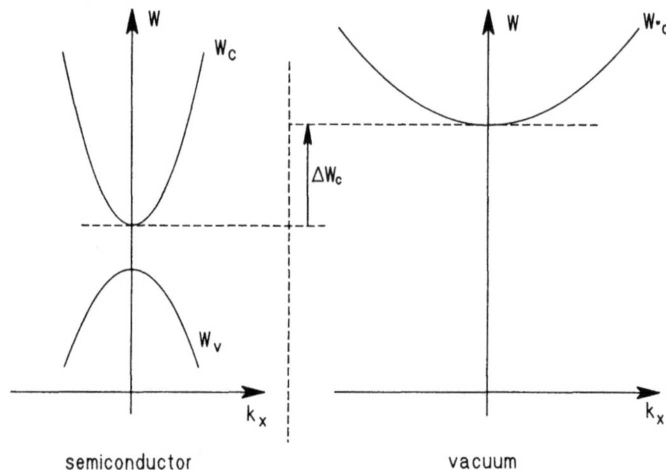

Fig. 3.1 Energy band structure in semiconductor and vacuum

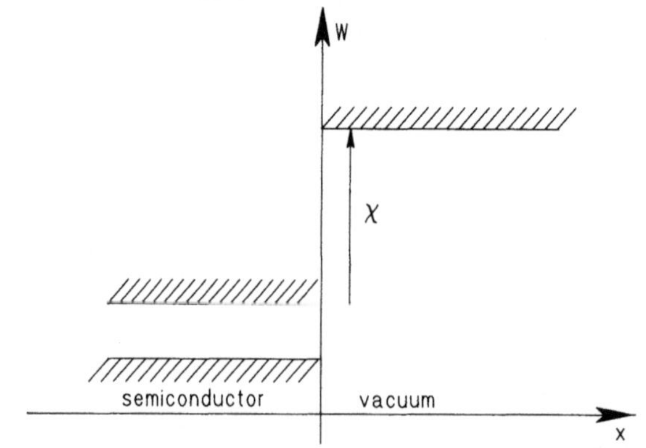

Fig. 3.2 Energy bands at the semiconductor-vacuum interface

is indicated by ΔW_c in the figure. In the present case of a semiconductor-vacuum interface, ΔW_c is usually called the *electron affinity* χ. Comprehensive tables of the electron affinity χ and the bandgap W_g for a number of semiconducting materials can be found in [28]. Typical values for χ are e.g. 4.01 eV for Si, 4.07 eV for GaAs, and 3.5 eV for AlAs.

In essence, Fig. 3.1 consists of two pictures, showing the energy dispersion in each material's own k-space. Now we change our point of view and look at the interface in position space. Figure 3.2 shows the energy bands as they change abruptly from the semiconductor bands to the vacuum band at the interface. (A possible bending of the bands is not recognizable on the length scale of the figure.) Of course, in this picture the k-dependence of the energy

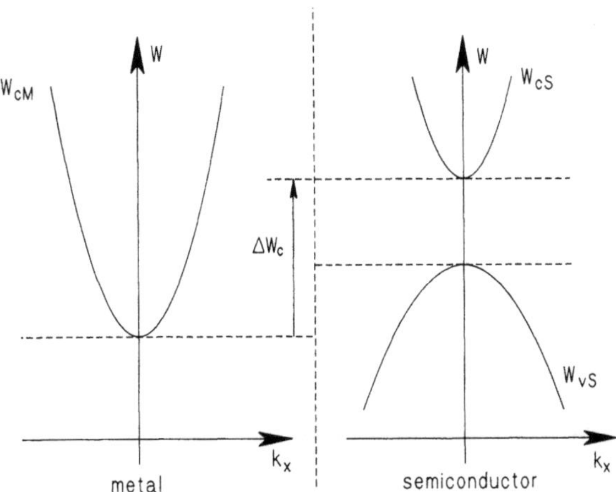

Fig. 3.3 Energy band structure in semiconductor and metal

is no longer visible; only the band edges are shown. The allowed energy ranges are still indicated in the figure by hatching. This kind of figure is called "energy band model". In the representation of Fig. 3.2, the energy barrier that prevents the electrons in the semiconductor from escaping into the vacuum becomes obvious.

As another example, we consider the case of a metal-semiconductor interface. The basic configuration of the energy bands in this kind of interface is depicted in Fig. 3.3. For simplicity, only simple parabolic $W(\mathbf{k})$-relationships are shown for each band in the figure, since this suffices to explain the basic behaviour of electrons at the interface. The essential features of Fig. 3.3 are the presence of allowed states in the metal throughout the energy range where the semiconductor band structure has its valence band maximum, energy gap, and conduction band minimum, i.e. where the electron transport processes take place. The actual band structures of semiconductors and metals usually look more complicated (see for example [8], [25]). Especially, the distinction between conduction and valence bands in a metal is blurred since the bands in a metal overlap each other, and a distinction between completely filled and empty bands at $T = 0$ is not possible. However, the energy dispersion relations for many metals are very similar to that of free electrons, except for k-vectors near the edges of the Brillouin zone, where Bragg reflections occur and the band structure is distorted [8], [25]. Thus, in simulations often the free electron mass is taken as the metal effective mass (see e.g. [29]).

The relative positions of the band structures on the energy scale are defined again by the band offset ΔW_c, as indicated in Fig. 3.3. As will be explained in Chapter 6, ΔW_c can be transformed into the barrier height, i.e. the difference between the metal Fermi level and the semiconductor band edge, which is used in practice for metal-semiconductor contacts as an indicator of the band offset.

As a result of the preceding discussion, we state that the primary parameter of the interface is the displacement of the energy band structures of the two material species on the energy axis, expressed by the band offset ΔW_c. All other energy relations result from this parameter together with the involved band structures.

3.2 Interface States

In this section, we are going to look in more detail into the situation at the interface. At the same time we will refine the so far coarse interface model. Our hitherto assumption of an abrupt transition of the band structure is an idealization, a first approach to the description of the interface. In reality, we find a certain transition region near the interface, where the energy band structure is not identical to either of the two bulk structures. The reason is that the crystal structure of the involved materials changes at the interface. Displacements and distortions of the crystal atoms occur in the vicinity of the interface, which have their effects on the energy band structure [27].

The most important feature of the energetic structure at a semiconductor interface is the occurrence of additional interface states in the energy gap [27]. One can view these states as a consequence of the perturbation of the bulk crystal caused by the formation of the surface [27].

Mainly two basic reasons exist for the presence of interface states. These are

1. disarrangements of the ideal periodic crystal lattice,
2. contamination of the interface.

The first reason is the existence of the interface itself. The concept of the energy band structure is based on the idealization of a periodic, infinitely extended crystal. It is only in this ideal case that the eigenstate of a definite energy W is a Bloch function with the quantum number $\mathbf{k}$, i.e. the wave number. Hence, also the energy dispersion relation $W(\mathbf{k})$, which gives rise to the allowed bands and the forbidden bandgaps, is clearly defined only in an infinite crystal. Already the simple ending of the crystallic structure, i.e. the semiconductor-vacuum interface, represents a drastic violation of this assumption of periodicity. Thus it is no wonder that in the vicinity of the interface deviations of the bulk band structure occur. Further, the forces on the atoms and thus their equilibrium positions are different near the interface than in the infinite crystal, because the balance of forces from all directions in the infinite crystal is perturbed at the interface where atoms on one side of the interface are missing. This adds to the displacements and variations of the energy band structure at the interface.

The various possible interface states that can arise from this situation are

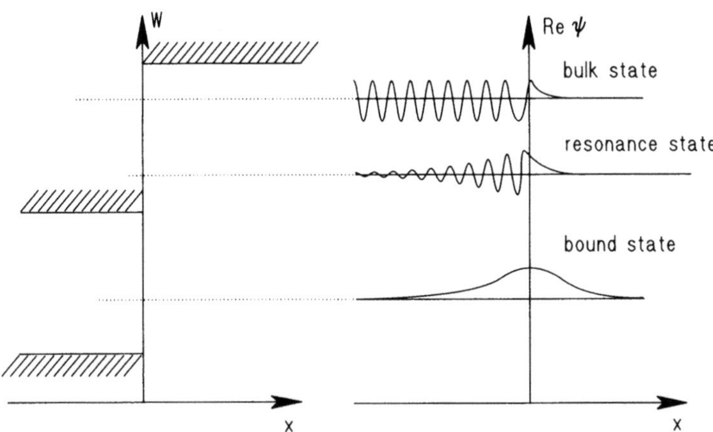

Fig. 3.4 The different kinds of interface states

depicted in Fig. 3.4 [27]. The left part of the figure shows the energy band model of a semiconductor-insulator interface, while the right part displays the basic behaviour of the wave functions at the respective energies, as indicated. The uppermost, so-called "bulk state" oscillates periodically and extends throughout the semiconductor. Since the energy of this state lies in the forbidden band of the insulator, it decays exponentially as we move away from the interface into the insulator. From the viewpoint of the semiconductor, this state is a normal conduction band electron; from the viewpoint of the insulator, however, an additional state has been created in the bandgap that is not present in the insulator bulk, but is located at the interface. A pictorial interpretation of this state is an electron travelling in the conduction band of the semiconductor and, when encountering the interface, tunnels into the insulator for a while before it is reflected back to the semiconductor.

Because of the leakage of electrons from the conduction band into the neighbouring bandgap, this kind of state is also referred to as "induced gap state". To be clear, such states do not only occur in semiconductor-insulator interfaces. They are present whenever a band of allowed energies is separated by the interface from a bandgap. They are especially important in metal-semiconductor contacts, where the electrons tunnel from the metal into the bandgap of the semiconductor. This type of state in particular is known as "metal-induced gap states" (MIGS) [27, 30].

Two other kinds of state are shown in the figure. The probability density of the so-called "bound state" has a maximum at the position of the interface, and decreases exponentially away from the interface in both materials. The probability of the "resonance state" also decreases in both directions, but oscillates at the same time in the semiconductor, i.e. shows the behaviour of a damped oscillation. The two latter states have zero ampli-

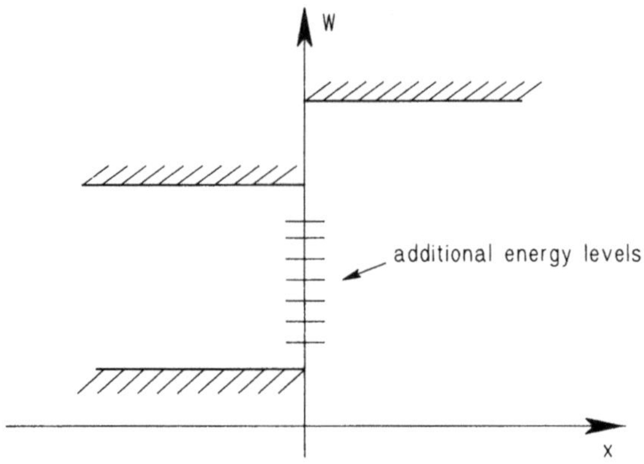

Fig. 3.5 Impurity atoms at the interface

tude far away from the interface, and thus describe an electron that is located at the interface. They have no analogy neither in the bulk semiconductor nor in the insulator; hence they represent true additional interface states.

The second cause for interface states are impurities at the interface. If for instance the two material layers that form the interface are fabricated in separate manufacturing steps, unwanted impurity atoms have a chance to contaminate the surface of the first layer before the second layer is deposited. Even if utmost cleanliness is maintained, impurity atoms that are present anyway in the involved materials can diffuse through the crystal during subsequent high temperature process steps until they reach the interface. Because of the strong perturbation of the crystal lattice at the interface—the number of vacancies or dislocations at this place can be very high—the atoms have an increased probability to stop their diffusion movement and to settle down at the interface (segregation). Whatever the cause is, we have to take into account the possibility of an enhanced concentration of impurities at the interface. As indicated in Fig. 3.5, these impurity atoms create additional localized energy levels in the bandgap at the interface. The impurities have an effect like a local doping of the semiconductor at the interface. They act as donors or acceptors, and represent centers for the recombination of electrons and holes (traps).

3.3 Surface Charge and Dipole Densities

The interface states give rise to positive as well as negative electric charges in the immediate vicinity of the interface. These charges exist in a layer that extends somewhat above and below the interface and has a finite but relatively small thickness. Usually, the thickness of these layers is very much

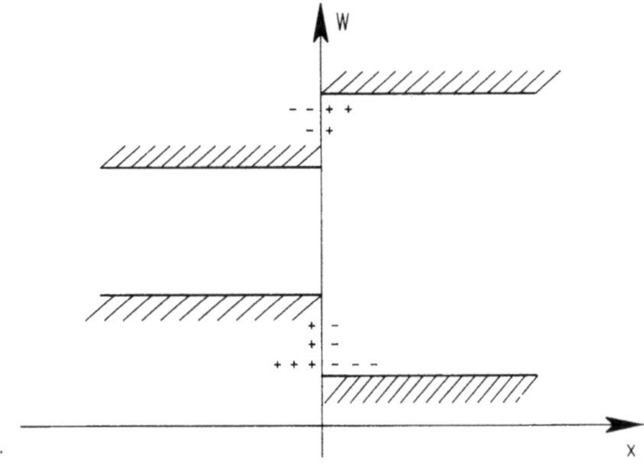

Fig. 3.6 Induced gap states

smaller than the characteristic lengths of our devices. Thus, in a simulation of the device, we normally make an abstraction from the finite interface layer to an ideal two-dimensional surface. In other words, we constrict the layer to zero thickness. The charge distributions in the interface layer then degenerate to surface charge densities and surface dipole densities.

An important role in the formation of surface dipoles is played by the induced gap states [27]. In order to understand this, we consider Fig. 3.6. (The following discussion is an extract of Tersoff's theory of heterojunction band lineups [31, 32].) As mentioned above during the discussion of the bulk-like interface states, electrons can tunnel from the valence band on the left across the interface into the bandgap on the right. This movement leads to an excess negative charge on the right. Since the tunneling electrons leave the neutral semiconductor on the left, their absence creates positively charged holes in that material's valence band [33].

A charge transfer is also associated with the tunneling of states at the conduction band step. However, this effect is somewhat more difficult to understand. (The mechanism is explained in more detail by Tersoff in [31].) Electrons that are in the conduction band on the left can tunnel into the bandgap on the right. But most of the states in the conduction band of a (non-degenerate) semiconductor are unoccupied. How can these states give rise to charge transfer across the interface? The reason is that even the pure existence of the conduction band states, regardless of their occupancy, influences the number of states in the valence band. According to a "sum rule", stating that the total number of available states in a semiconductor is approximately position independent [31, 33, 34], the presence of the additional induced states in the bandgap near the conduction band diminishes the number of available states in the allowed bands. Thus a valence state, which would be occupied in the bulk, is "replaced" near the interface by a state in the bandgap because of the sum rule. If the interface state is unoccupied, a net positive charge is the consequence because fewer states are occupied at the interface compared to the neutral bulk. Although this charge occurs in fact as a hole in the valence band, it is a "book-keeping convention" [31] to count them at the place of the empty gap states, as in Fig. 3.6.

Summarily, we see from Fig. 3.6 that positive and negative charge clouds can result from the existence of induced gap states. The centers of the negative and the positive charge lie a small distance apart; hence a dipole moment [35] at the interface is the consequence. In Fig. 3.6, we see one dipole near the valence band and one associated with the conduction band. In general, these two dipoles to not cancel. The net interface dipole depends on the detailed band structure of the materials and the sign of the discontinuities.

As mentioned at the beginning of this section, we want to conceptually idealize the charges inside the interface layer to corresponding *surface densities*. Formally, this can be done as follows. Looking into the interface layer on a microscopic level, we see a space charge distribution ρ, which might appear for example like the one shown in Fig. 3.7. On a macroscopic level, the interface layer degenerates to a surface of zero thickness; the charges in the interface layer then are located on the surface and are described by the surface charge density σ and the surface dipole density τ.

The surface charge density is obtained from the interface charge (Fig. 3.7) simply by collecting all the charge that belongs to the actual surface element:

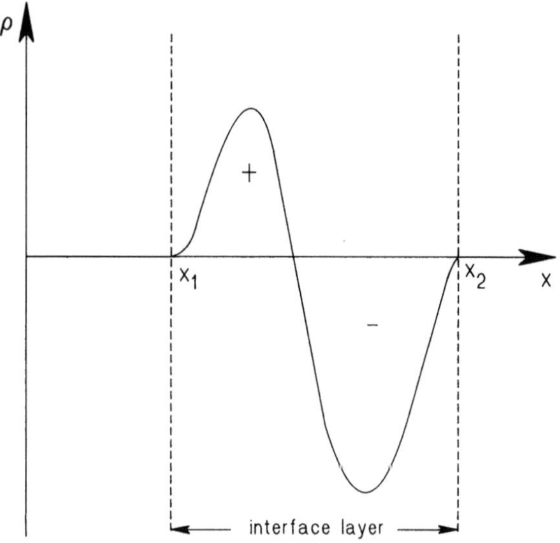

Fig. 3.7 A possible distribution of interface charges

$$\sigma(y, z) = \int_{x_1}^{x_2} dx\, \rho(x, y, z). \tag{3.1}$$

By this procedure, the surface charge density comes out as the total charge on the surface element, divided by the area of that element. Thus the natural concept of a surface charge density is preserved. It is assumed in (3.1) that the local coordinate system is such that the x-axis is normal to the interface, while y and z are the surface coordinates. Hence, the vector (y, z) denotes the position on the surface, and the integral with respect to x collects all charge belonging to the surface element at this position.

If the total numbers of positive and negative charges in the interface layer are equal, they cancel when computing the surface charge, and $\sigma = 0$ is the consequence. However, also in this case an electrostatic effect comes from the interface charges if the centers of gravity of negative and positive charges do not coincide. This effect is preferably described by introducing a surface dipole density τ that is defined by

$$\tau(y, z) = -\int_{x_1}^{x_2} dx \int_{x_1}^{x} dx'\, \rho(x', y, z). \tag{3.2}$$

One can think of a surface dipole density as the sum of all dipoles on a surface element, divided by the area of that element [35]. This can be computed by first combining two individual opposite point charges into a dipole moment (i.e. absolute charge times distance vector [23, 35]), and second by summing up the normal components of all the dipoles in the interface layer. One easily verifies that the total dipole moment is invariant with respect to permutation of two equal point charges between the pairs. As a result, we find an alternative definition of the surface dipole density [36]:

$$\tau_{\text{alt}}(y, z) = \int_{x_1}^{x_2} dx\, x \cdot \rho(x, y, z). \tag{3.3}$$

It can be shown by applying partial integration to the right-hand side of (3.3) that this expression is identical to the first definition (3.2) plus a term that is proportional to the surface charge σ . The additional term comes from including in the integral those charges for which a partner to form a dipole with is missing. Hence the two definitions are equivalent if the surface element is a pure dipole, i.e. contains no net charge ($\sigma = 0$). However, Eq. (3.2) is the definition to be preferred since it will be shown below that it directly relates to the potential discontinuity at the interface.

Next, we discuss the effect of surface charge density and surface dipole density on the electric field and the electrostatic potential, since this will lead us directly to the interface conditions for the Poisson equation.

In order to investigate the effect of the surface charge, we integrate Maxwell's third equation (2.26)

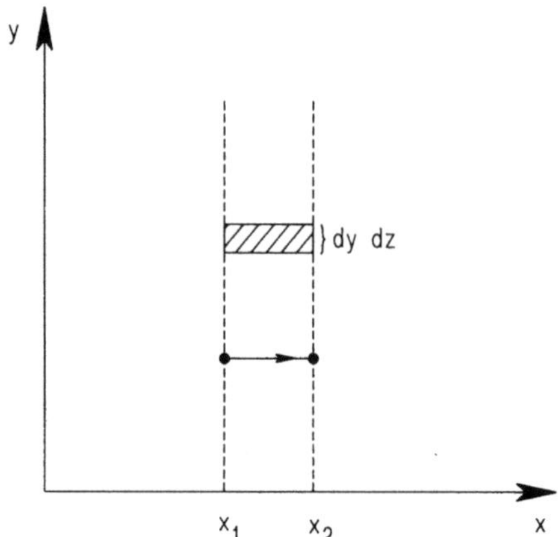

Fig. 3.8 Interface layer and integration volume

$$\text{div } \mathbf{D} = \rho \tag{3.4}$$

over a small cylindrical volume depicted in Fig. 3.8, which lies inside the interface layer and is bounded by a surface area $dy\,dz$. Application of Gauss' law yields [23]

$$D_x(x_2) - D_x(x_1) = \sigma, \tag{3.5}$$

where σ is again the surface charge density (3.1). Thus, we find that the effect of the surface charge is a discontinuity of the normal component (the x-component in the present case) of the displacement vector $\mathbf{D}$.

The effect of the surface dipole density is found by integrating the relation between the electrostatic potential φ and the electric field $\mathbf{E}$ (2.28) across the interface layer along a path that is also indicated in Fig. 3.8:

$$\varphi(x_2) - \varphi(x_1) = -\int_{x_1}^{x_2} dx\, E_x(x). \tag{3.6}$$

The electric field inside the interface $E_x(x)$ can be obtained by the same consideration that led to (3.5), with x_2 replaced by x:

$$E_x(x) = E_x(x_1) + \frac{1}{\varepsilon}\int_{x_1}^{x} dx'\, \rho(x'). \tag{3.7}$$

In (3.7), we switched from $\mathbf{D}$ to $\mathbf{E}$ with the help of $\mathbf{D} = \varepsilon\mathbf{E}$ (Eq. (2.27)). Inserting (3.7) into (3.6), we find

$$\varphi(x_2) - \varphi(x_1) = -\frac{1}{\varepsilon}\int_{x_1}^{x_2} dx \int_{x_1}^{x} dx'\, \rho(x')$$

$$- E_x(x_1)\cdot(x_2 - x_1), \tag{3.8}$$

which gives us the discontinuity of the electrostatic potential in terms of the interface charge. Passing to the notation of the macroscopic view of an infinitesimally thin interface, we find with (3.2) in the limit $x_1 \rightarrow x_2$ that the electrostatic effect of the surface dipole density is a discontinuity of the potential at the interface [35],

$$\varphi_2 - \varphi_1 = \frac{\tau}{\varepsilon}, \qquad (3.9)$$

where φ_μ means the potential on side μ of the interface. Through this effect the surface dipoles cause an additional shift of the energy bands at the interface, which influences the value of the band discontinuity ΔW_c. (It is somewhat problematic to tell exactly what the dielectric constant in (3.9) should be, since the integrations are on the length scale of the atomic distance, where the macroscopic concept of relative permittivity looses its validity. However, since we do not draw any quantitative conclusions from (3.9), this question is not addressed further.)

3.4 Electrostatic Interface Conditions

In conclusion of the preceding sections, we state that primarily the semi-conductor interface is described by a difference between the conduction band edges ΔW_c. This concept has to be complemented by the consideration of localized states at the interface. The interface states can be occupied by electrons, which in turn can lead to surface charges, surface dipoles, and surface recombination at the location of the interface.

Thus, we are able to characterize the *electrostatic* properties of the interface by

- ΔW_c [VAs]—conduction band edge discontinuity;
- σ [As m^{-2}]—surface charge density.

The interface dipole density does not appear in this list because—as we will see below—its effect is absorbed in the band edge discontinuity. Interface recombination is not regarded as an electrostatic effect; hence it will be discussed in the following chapters when we turn to the transport effects at various interfaces.

It will be assumed in the following that the characteristic parameters ΔW_c and σ are known for the interface under investigation. We have seen that the lineup of the energy bands and the distribution of interface states are determined by the interaction of the two crystal structures in contact at the interface. Thus the interface parameters cannot be composed from separate attributes of the involved materials alone. They have to be determined from experiments and/or theoretical considerations for each particular interface anew. The knowledge thus obtained can be either a simple numeric value or a more or less sophisticated model that allows the parameter to depend on the actual electronic state of the interface. For example, the band discontinuity (or the barrier height) of a metal-semiconductor contact may depend on the electric field at the interface, or the surface charge density at a semiconductor-insulator interface on the local electron quasi-Fermi level. A large number of papers can be found in the literature that deal with the theoretical and experimental investigation of band offset and interface states at semiconductor interfaces. However, because we are mainly concerned with the simulation of semiconductor devices in this book, we are in a sense only users of such models. Thus, we content ourselves here with a presentation of the most important models, but do not immerse too deep in this area of research.

The most simple and trivial case of model is a constant value of a parameter for a particular interface. For instance, in the simulation of heterojunction devices, the conduction band edge discontinuity between Si and SiO_2 is often taken to be a constant in the range 3.0–3.2 eV (e.g. [29, 37, 38]).

A more complicated model for the same parameter—also at the Si-SiO_2 interface—is the consideration of tunneling or image force effects by assuming a model equation that relates ΔW_c to the local electric field or potential [39, 40, 41].

A simple model equation for the surface charge density σ can be derived from the density of interface states in the bandgap (traps). The interface state density $D_t(W_t)$ specifies the number of energy levels on a given interface area and in an infinitesimal energy interval around the energy W_t; it is measured in units of $cm^{-2}eV^{-1}$. A simple model for σ then is [42]

$$\sigma = -qD_0(\phi_n - \phi_0), \tag{3.10}$$

where ϕ_n is the quasi-Fermi energy at the interface. The constant parameters of this model are the interface state density D_0 (assumed energy independent) and the "neutral Fermi level" ϕ_0. The model assumes the presence of a constant number D_0 of interface states throughout the bandgap. Typical values for D_0 at Si-SiO_2 interfaces are in the range of 10^9–10^{12} $cm^{-2}eV^{-1}$ (see e.g. [43] and references therein). The interface states are occupied or unoccupied according to their relative position to the quasi-Fermi energy. Since we can have a mixture of donor and acceptor levels, the model allows to specify a quasi-Fermi energy ϕ_0 where the interface states as a whole appear neutral. In addition to these rechargable states, *fixed* interface charges also have to be taken into account [44].

Outgoing from the characteristic parameters of the interface, we are now able to state the electrostatic interface conditions, i.e. the interface conditions for the Poisson equation. The Poisson equation is a second order differential equation for the electrostatic potential φ. According to Section 2.5, we need two conditions at a semiconductor interface that connect in some way the potential and its derivative in the two regions separated

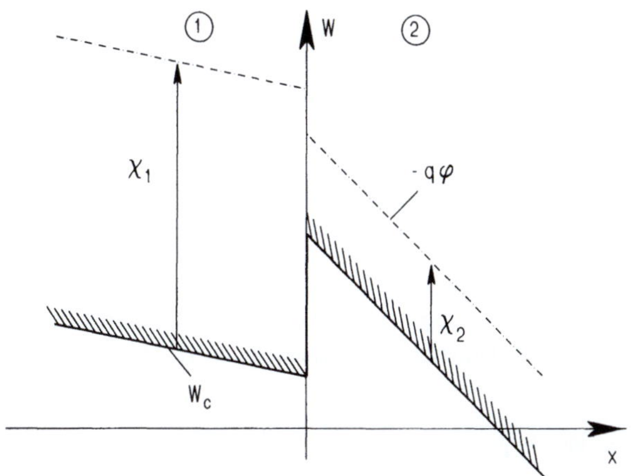

Fig. 3.9 Energy bands and electrostatic potential at the interface

by the interface. Following the reasoning so far, we write the interface conditions for the Poisson equation in the form

$$W_{c2} - W_{c1} = \Delta W_c \qquad (3.11)$$
$$D_{2x} - D_{1x} = \sigma. \qquad (3.12)$$

It is assumed in these conditions (as throughout the book) that the x-axis is normal to the local interface plane. Other orientations can be handled by appropriate coordinate transformations. As depicted in Fig. 3.9, the interface separates two regions, numbered 1 and 2, where region 1 is on the side of lower x-values and region 2 on the side of higher x-values. Physical quantities that are (mathematically) defined at the same point on the interface, but belong infinitesimally to one or the other side, are denoted by the respective index of the region.

Eq. (3.11) specifies the energy band shift of the two regions by the interface parameter ΔW_c. Eq. (3.12) determines the jump of the normal component of the electric displacement as a function of the interface charge density σ. (We already found (3.12) in Section 3.3 in the discussion of the effect of the surface charge, which has led us to Eq. (3.5).)

If we are dealing with a *boundary* instead of an interface, only one condition can be specified according to the discussion in Section 2.5. Normally one chooses one or the other of Eqs. (3.11) and (3.12), depending on the type of interface. In special cases, also a linear combination of the two interface conditions is used. We shall discuss this point in subsequent chapters when we address the different interface types specifically.

Since the above formulation of the interface conditions does not contain the electrostatic potential, we shall now rewrite the equations in order to clarify the relation to the Poisson equation as an equation for φ. The connection of the interface conditions to the electrostatic potential φ is as follows. The potential φ is related to the x-component of the displacement $\mathbf{D}$ in (3.12) by (2.27) and (2.28) as

$$D_{ix} = \varepsilon_i E_{ix} = -\varepsilon_i \frac{\partial \varphi_i}{\partial x}. \tag{3.13}$$

Regarding (3.11), we note that Fig. 3.9 shows that the difference between the "macro potential" [45] $-q\varphi$ and the conduction band edge is the electron affinity χ. Hence, the relation between φ and W_c for a certain material is given by

$$-q\varphi = W_c + \chi. \tag{3.14}$$

The idea behind this is that the electron would have a potential energy $-q\varphi$ if the crystal was not present; the electron affinity expresses the additional binding forces that the crystal atoms exert on the electron [45]. For this reason, the electron affinity is a property of the material. With (3.13) and (3.14) the interface conditions expressed for the potential become

$$-q(\varphi_2 - \varphi_1) = \chi_2 - \chi_1 + \Delta W_c \tag{3.15}$$

$$\varepsilon_2 \frac{\partial \varphi_2}{\partial x} - \varepsilon_1 \frac{\partial \varphi_1}{\partial x} = -\sigma. \tag{3.16}$$

In the early days of semiconductor heterojunction physics it was common to assume that the potential φ is continuous at the interface [28]. Then from (3.15) follows immediately

$$\Delta W_c = \chi_1 - \chi_2. \tag{3.17}$$

This equation is known as the "electron affinity rule". It is usually attributed to Anderson, although in the original paper [46] Anderson only notes (in connection with a respective figure) that "voltage is continuous in the absence of dipole layers", while he considered ΔW_c as an independent parameter which he determined from experiments.

If (3.17) would be valid, the conduction band edge discontinuity could be determined for arbitrary interfaces once the electron affinities of the two involved materials are known. This would simplify the situation considerably, because from the knowledge of the material parameters χ alone the interface parameter ΔW_c for all possible combinations can be determined. Unfortunately, numerous experiments in the following years have shown that *the electron affinity rule is not valid in general* (see [47] for a review). From our discussion in Section 3.3 we understand why this is the case. Eq. (3.15) shows that (3.17) is only valid if $\varphi_2 - \varphi_1 = 0$. However, in (3.9) we found that the potential difference is proportional to the surface dipole density τ. In Section 3.3 we saw that in general a surface dipole has to be expected because of the non-coincidence of positive and negative interface charges. Hence, zero dipole density rather is the exception than the normal case, and the electron affinity rule is only valid for those special material combinations where the dipole density is negligible.

By solving (3.15) for ΔW_c and inserting (3.9), we arrive at

$$\Delta W_c = \chi_1 - \chi_2 - \frac{q\tau}{\varepsilon} \tag{3.18}$$

as a possible model equation for the band discontinuity, which requires knowledge of the electron affinities *and* the surface dipole density. However, since the surface dipole normally is not known separately, it is convenient to choose ΔW_c directly as the interface parameter that can be determined experimentally and that characterizes the band lineup. If we had a practicable theory that gives us the interface dipole density as a function of the materials, Eq. (3.18) shows the way to predict the conduction band edge discontinuity for arbitrary interfaces.

Tersoff's theory is a promising candidate for a prediction of the band lineup [47]. In principle, he postulates the existence of a "neutrality level" W_N in each material species, such that the interface dipole is given by the respective difference

$$\tau = \frac{\alpha\varepsilon}{q}(W_{N2} - W_{N1}). \tag{3.19}$$

The susceptibility α is a proportionality coefficient. The neutrality level lies by an amount of W_B above the valence band edge,

$$W_N = W_v + W_B, \tag{3.20}$$

where W_B is a material parameter. Tersoff suggests for semiconductors that W_B is the energy of the branch point of the band structure, i.e. the energy where the imaginary energy dispersion in the bandgap branches into the conduction and valence bands [31]. For metals, W_N is expected to be the Fermi energy [31].

Insertion of (3.19) and (3.20) into (3.18) and solving for ΔW_c yields

$$\Delta W_c = -\frac{\Delta\chi}{1+\alpha} + \frac{\alpha}{1+\alpha}(\Delta W_g - \Delta W_B), \tag{3.21}$$

where Δ denotes the differences of the respective quantities between material 2 and 1; the valence band step has been substituted with $\Delta W_v = \Delta W_c - \Delta W_g$ by the discontinuities of the conduction band edge and the bandgap W_g. Tersoff further argues that α is large; hence the conduction band step becomes

$$\Delta W_c = \Delta W_g - \Delta W_B \tag{3.22}$$

in a zero-order approximation. Within this theory, knowledge of the bandgap W_g and "midgap energy" W_B of the involved materials suffices to predict the band discontinuities. Respective tables of material parameters can be found e.g. in [47].

Many developers of device simulators (sometimes tacitly) choose the potential origin such that the electrostatic potential φ coincides with the "intrinsic Fermi potential" φ_i [48]. The intrinsic Fermi potential is given by

$$\varphi_i = -\frac{1}{q} W_i, \tag{3.23}$$

where W_i is the "intrinsic Fermi energy" defined by

$$W_i = W_c - \frac{W_g}{2} + \frac{k_B T}{2} \ln\left(\frac{N_v}{N_c}\right). \tag{3.24}$$

In this equation, W_g is again the bandgap of the semiconductor, and N_c and N_v are the effective densities of states of the conduction and valence band, respectively. These quantities are material parameters of the considered semiconductor. Pictorially, W_i lies near the middle of the bandgap; it is the energy level of the Fermi energy of the intrinsic (i.e. undoped) semiconductor in equilibrium.

As has been pointed out by Marshak [49], the use of the intrinsic level for the electrostatic potential is appropriate only in the case of homogeneous device regions, because W_i is not parallel to $-q\varphi$ if χ, W_g, N_c, or N_v are position dependent.

Provided that the use of W_i is valid, we can rewrite the boundary condition (3.15) for the intrinsic potential. If we eliminate W_c in (3.24) with the help of (3.14), solve for φ and insert into (3.15), we arrive at the respective interface condition:

$$\varphi_{i2} - \varphi_{i1} = -\frac{1}{q}\Delta W_c + \frac{1}{2q}(W_{g2} - W_{g1}) - \frac{k_B T}{2q} \ln\left(\frac{N_{v2} N_{c1}}{N_{c2} N_{v1}}\right). \tag{3.25}$$

This equation expresses the band discontinuity in terms of φ_i; the step in φ_i at the interface is determined by the interface parameter ΔW_c and the material parameters.

Charge Transport Across the Interface   4

4.1 Full Quantum Transport

Quantum-mechanical calculations of electron transport in semiconductor devices can be based on several formulations, among which are eigenfunctions of the Schrödinger equation, the density matrix [3], an approximate diagonalized density matrix, or the Wigner function [26]. In any case, boundary conditions at the edges of the device and interface conditions at material interfaces have to be considered. In the following, we will briefly discuss the basic concepts for quantum-mechanical boundary and interface conditions. These considerations will have implications on the respective conditions in the semi-classical case, which will be described in the subsequent sections.

4.1.1 Boundary Conditions

In quantum-mechanical textbooks, electron transport is usually studied for idealized systems, which are in most cases either infinitely extended or surrounded by infinitely high potential walls. Common boundary conditions assumed for these cases are plane waves in an asymptotically flat potential far away from the active region, or vanishing wave functions at the infinite wall. A practical electron device, however, is normally bounded by insulating layers and metallic contacts, which connect the device to a voltage source or an electric circuit.

Since the potential steps occurring at boundaries adjacent to good insulators usually have heights of several electron volts (see Chapter 5), they can be reasonably well modelled with respect to electron transport as being infinitely high. This means that the boundary is treated as ideally reflecting, which can be expressed as the condition of a vanishing wave function.

Metallic contacts, on the contrary, require a more detailed consideration. Contacts maintain the connection of the device to an external circuit, which in the simplest case is a voltage source. Hence, the externally controlled parameter is the applied voltage, and the concept of an incident electron wave, as implied by the plane wave boundary condition in the textbook examples mentioned above, cannot be used. The problem of metal contacts on quantum devices has been discussed in detail very clearly by Frensley [50].

The contacts with their high electron densities and scattering rates represent complicated systems themselves. However, we are mainly interested in the electronic behaviour of the device, while the contacts serve only for the application of voltage and the supply of current. For this reason, we usually place the boundary of the domain of interest at the metal-semiconductor interface, and consider the metal layer as part of the environment (see also [51]). According to Section 2.5, this means that we need a simplified model of the metal that suffices to describe the interaction with the semiconductor reasonably well.

The metal is considered as an electron reservoir [50]. This means that the metal electrons are assumed to be in thermodynamic equilibrium at a given temperature and Fermi energy. Because of the high scattering rates, the electrons exchange energy very frequently with each other and thus take on an equilibrium state. The Fermi level is determined by the electric circuit connected to the contact, according to Kirchhoff's laws. In the simplest case of a battery connected to a pair of contacts, the Fermi level is prescribed by the battery voltage. It should be emphasized that it is the difference of the *Fermi level* and *not the electrostatic potential* φ that corresponds to the voltage which is measured between two contacts [45]. Hence the difference between the Fermi level at a contact in equilibrium and in non-equilibrium is equal to the potential applied to the contact. If *we define the origin of the energy scale at the equilibrium Fermi level*, the metal Fermi energy corresponds directly to the applied potential. This definition will be assumed throughout the book.

The concept of the contact as a reservoir involves the idea that the contact is something macroscopic; i.e. the scattering rate is so high that, despite of currents flowing, the equilibrium distribution with constant Fermi level is approximately preserved throughout the metallic region. Because of this macroscopic property of the reservoir, irreversibility is introduced into the quantum mechanical system. For this reason, the state of the system "reservoir plus device" cannot be described by pure quantum states. Instead, a statistical representation of mixed states [2] has to be invoked [50]. This can be a density matrix formulation [2], or the Wigner distribution introduced in Chapter 2.

As discussed in Chapter 2, the Wigner distribution is advantageous with respect to simulation of quantum devices. Hence, the reservoir is preferably described by an equilibrium distribution function $f^0(\mathbf{x}, \mathbf{k})$ [50]. In principle, the Wigner functions in the contact and the device as part of a whole quantum-mechanical system are connected by the non-local quantum interaction (2.3) in the Wigner equation (2.2). However, by introducing the boundary at the metal interface and assuming a constant equilibrium distribution in the metal, the contact is placed outside the system, and the interaction between metal and device is reduced to a local boundary condition.

It has been pointed out in Section 2.5.1 that from mathematical considerations boundary conditions for the Wigner equation must be specified only on those boundary portions where electrons flow *into* the device. Hence, for a metal boundary at $\mathbf{x} = \mathbf{x}_0$, where the device is at $x < x_0$ (a "left-side" boundary), the boundary condition for the Wigner equation reads [50]

$$f(\mathbf{x}_0, \mathbf{k}) = f^0(\mathbf{x}_0, \mathbf{k}) \qquad (v_x(\mathbf{k}) > 0). \tag{4.1}$$

The equilibrium distribution function on the right-hand side depends on the temperature and the Fermi level of the reservoir and thus introduces the effect of external voltages on the device. In general, f^0 is the Fermi-Dirac distribution. In simplified, one-dimensional analyses the integral of the Fermi-Dirac distribution over the transverse wave vectors is used instead [50]. The boundary condition (4.1) specifies the distribution function only on the inflow part of the boundary; on the outflow part the distribution function remains unspecified and is determined by the solution of the Wigner equation in the interior of the device. Since the inflow distribution is entirely independent of the outflow, the system is not symmetric with respect to time reversal; hence the boundary condition is irreversible. Frensley has shown that the irreversibility of the boundary assures assures the stability of the solutions of the Wigner equation. Implementations of the boundary condition (4.1) into numerical simulation codes have been reported in [50, 52, 53, 54].

The physical meaning of (4.1) is that the metal injects electrons into the device according to the equilibrium distribution in the contact. The electronic behaviour in the interior of the device as a reaction to this is determined by the Wigner equation. Electrons reaching the boundary from the interior are collected by the contact. Possible quantum-mechanical reflections at the interface are described in the framework of the Wigner equation by the non-local potential interaction inside the device and are not part of the boundary condition.

In a density matrix approach, the distinction of in- and outflowing electrons at the boundary is not as easy as in the Wigner representation [3], since the density matrix depends either on a pair of position coordinates, wave vectors, or energy quantum numbers, but not on position and wave vector simultaneously. Hence it is difficult to formulate the inflow boundary condition in this representation. In a one-dimensional treatment, this problem can be cirumvented because in that case we have exactly two contacts, and the current flow of each energy state in the density matrix has one of two possible directions. Hence, the states having particle current to the right can be associated with the left-hand contact and vice versa. The probabilities of the energy states can then be set according to the distribution in the associated contact [51]. Thus in one dimension it is not necessary to evaluate the flow at a definite location, i.e. the boundary. However, this approach does not seem to be extensible to higher dimensions.

While (4.1) represents the boundary condition for the electron transport equation, for a self-consistent simulation (i.e. the potential depends on the space charge because of the Poisson equation), we still need a boundary condition for the Poisson equation. As will be discussed in more detail in Section 6.4, the boundary condition for the Poisson equation at an ideal ohmic contact is the requirement of charge neutrality. Some authors prefer to associate the neutrality condition with the transport equation and the applied voltage with the Poisson equation, respectively. In this picture, the external voltage source raises the electrostatic potential at the boundary, and the electron distribution reacts to this variation in order to maintain charge neutrality. I myself prefer the view that the Fermi level of the contact metal is determined by the voltage source (this coincides with the respective discussion in [45], see above), and the potential reacts as to restore charge neutrality. The latter concept arises naturally as a limiting case from a detailed consideration of non-ideal contacts (see Chapter 6). However, in the end it is a matter of taste, because the effects are deeply interrelated, and it is idle to ask what is the effect and what is the cause[1].

At this point, it is interesting to note the similarity of concepts related to boundary conditions for quantum devices that are connected in electrical or in quantum waveguide circuits [55]. In the latter case, the devices are connected by waveguides that lead the electron wave without destroying the coherence, in contrast to the metallic wires in an electrical network. Lent and Kirkner [56] describe a numerical solution of the Schrödinger equation for a quantum device connected to electron waveguides ("leads"). (Note that for electron waves even an arc in a waveguide represents a "device" with a non-trivial transfer function [56].) Similar to the above discussion for metallic contacts, the leads are placed outside the simulation domain, and corresponding boundary conditions are formulated at the interfaces. Here the complex amplitudes of the incoming waves must be known, and the outgoing waves are determined by the interior of the device, similar to the treatment of incoming and outgoing electron flux discussed above. This leads to a boundary condition for the wave function ψ, which relates the incoming modes to the normal derivative of ψ and

---

[1] Nevertheless, differences in the numerical convergence behaviour might arise from different assignments of boundary conditions to volume equations in the implementation.

to the values of ψ at all other points along the boundary. Because of the wave-like nature of the device and its environment, pure quantum states have been considered. A direct transference of this boundary condition to the case of electric circuits appears difficult (though not impossible), because the concept would have to be extended to mixed states in order to be able to describe the reservoir characteristics of the contacts. The incoming waves then would have to be determined from the electron distribution in the metal.

4.1.2 Interface Conditions

The consideration of a material interface inside a quantum device appears to be much easier than the ohmic contact boundary condition, because in this case we have the same level of description, i.e. the microscopic quantum level, on both sides of the interface. This is in contrast to the boundary condition of the metallic contact, which required a macroscopic, dissipative description by equilibrium distributions on the metal side.

The behaviour of the Schrödinger wave function at a potential step has been studied in [2]; for heterojunctions see also [57] and [58]. If two-dimensional interface states can be ruled out on this microscopic quantum level, the requirement of particle conservation leads to

$$\frac{1}{m_1}\frac{\partial\psi_1}{\partial x} = \frac{1}{m_2}\frac{\partial\psi_2}{\partial x}, \tag{4.2}$$

$$\psi_1 = \psi_2, \tag{4.3}$$

as the interface condition for the wave function ψ. Here, the indices 1 and 2 refer to the respective values on either side of the interface; the x-axis is assumed as perpendicular to the local surface plane. In (4.2), we accounted for the possibility of different effective masses in the two regions separated by the interface [57, 59]. The problem of quantum interface conditions for nonparabolic bands has been addressed in [60, 61].

Some remarks are in order concerning these interface conditions. First, we note that the effective mass Schrödinger equation itself is already an approximation for electron propagation in infinite crystals, which is only valid near the band edges [62]. Hence, the concept that the effective mass changes its value abruptly at the interface may be questionable. Second, it might be doubtful to assume on one hand that the interface states have a finite size, while assuming on the other hand that the potential increases at the same time in a step-like manner (or at least over a distance much smaller than the electron wave length [2]). Thus, Eqs. (4.2)–(4.3) must be regarded as idealizations, which nevertheless can produce useful results [63].

Little can be found in the literature with respect to interface conditions for the Wigner equation. The problem is that the Wigner equation (2.2) is strictly valid only for a position-independent effective mass m. If m depends on position, it is not sufficient to use simply $m(\mathbf{x})$ in place of m in (2.2); rather, the Weyl transformation of the Schrödinger equation with position-dependent effective mass yields an inertia term with a complicated nonlocal functional of $m(\mathbf{x})$ [54]. Even in the case of an interface between homogeneous regions, where the effective mass is piecewise constant, this nonlocal dependency remains. Only sufficiently far away from the interface, the inertia term asymptotically approaches the form of (2.2) with the local constant mass. An interface condition for the full nonlocal effective mass Wigner equation [54] has not yet been elaborated.

A way to avoid these difficulties is to use (2.2) and to assume a constant effective mass througout the device, regardless of the actual local material composition [53, 64]. Alternatively, Frensley [50] proposed a semi-classical (local) approximation by placing $m(\mathbf{x})$ inside the gradient of the inertia term, if the mass discontinuity is small. In this case we may obtain an interface condition for the Wigner equation by a derivation similar to the treatment of the interface within the Schrödinger equation. Integration of (the modified) equation (2.2) along a short perpendicular path comprising the interface yields in the limit of vanishing path length

$$\frac{1}{m_1} f_1 = \frac{1}{m_2} f_2 \tag{4.4}$$

for parabolic bands, where m_1 and m_2 are again the respective effective masses, and f_1, f_2 are the Wigner functions at the interface on the different sides. As in the wave function representation, the interface condition expresses particle conservation at the interface. Condition (4.4) ensures that the normal current density, calculated from the Wigner function with (2.10) and (2.5), is continuous at the interface.

Since the Wigner equation is a differential equation of first order, one condition at an interface is necessary. This is in contrast to the Schröinger equation, which is of second order and needs the two interface conditions (4.2)–(4.3).

4.2 Semi-Quantum Interface Transport

4.2.1 Concepts

In this section, "semi-quantum transport" is considered. This means that quantum effects only play a role in the immediate vicinity of the interface, while otherwise in the bulk of the semiconductor the semi-classical Boltzmann equation is valid (cf. the related remarks in Section 2.2). Thus,

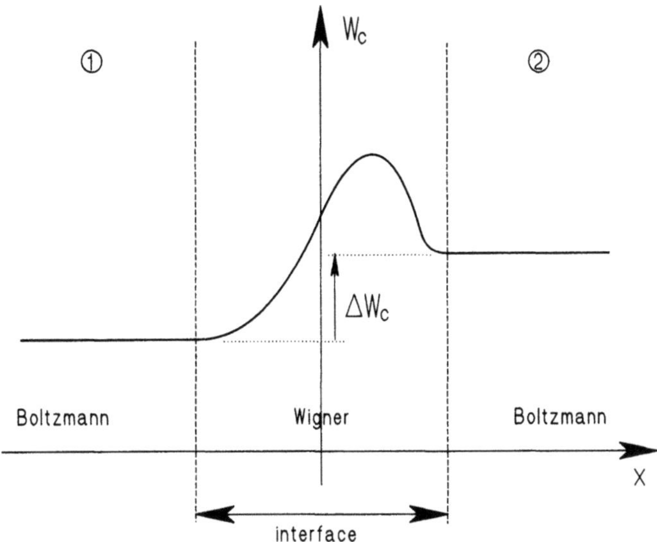

Fig. 4.1 Conduction band edge in the vicinity of the interface

the analysis is devoted to device classes where quantum-effects in the bulk
are not important, but occur solely near the interfaces and can be included
into the interface conditions.

It has been stated in Section 2.2 that the Boltzmann equation is only valid
if the electric field and hence the band edges do not change rapidly in space.
As mentioned in Chapter 3 and depicted schematically in Fig. 4.1, strong
variations of the band edges may occur inside the interface. Thus we expect
that there exists a thin layer around the metallurgical junction where the
variations are so strong that the Boltzmann equation is not valid in this
region, and quantum effects (i.e. the Wigner equation) have to be taken into
account. Outside this layer, the potential is smooth and well-behaved, and
semi-classical transport can be considered to be a valid description of the
transport effects in the bulk regions.

Let us take an electron with a definite energy. In steady state and in the
absence of scattering, the solution of the Boltzmann equation is zero everywhere in phase space except on the classical path. The actual distribution
function inside a device then is a superposition of many paths, i.e. a distribution of electrons with various energies.

If the kinetic energy of the electron is smaller than the top of the barrier
in Fig. 4.1, the electron is reflected at the barrier. Fig. 4.2 shows the two
possible paths of an electron with definite energy impinging on the barrier
from the left or the right, and being reflected. Apart from scattering, these
paths are described by the Boltzmann equation, and are thus valid outside
the interface region. Inside the interface, however, the Boltzmann equation

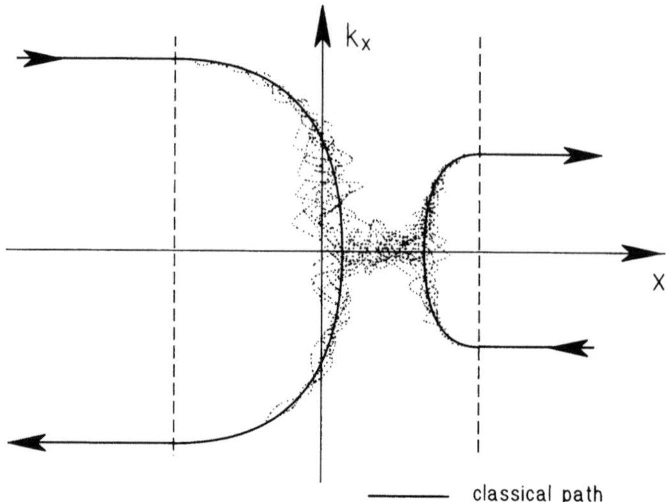

Fig. 4.2 Classical and quantum paths at low electron energies

looses its validity due to the strong variations of the band edge. This means that besides the single classically allowed path, other non-classical paths are possible with a certain probability[2]. Because of energy conservation in steady state, all these paths have to end up again in the classical path when leaving the interface region.

As indicated in the figure, some of the non-classical paths may cross the barrier. This means that, although classically forbidden, the electron can cross the barrier with a certain probability. This is the quantum-mechanical "tunnel effect".

A corresponding situation occurs if the kinetic energy is higher than the potential barrier. As shown in Fig. 4.3, the classical paths cross the barrier in this case. But in the quantum regime of the interface region, some of the reflecting paths have also non-zero probability, which is indicated in Fig. 4.3. This effect is called "quantum reflection".

In summary, contrary to the classical behaviour which allows only one of the possibilities of transmission or reflection, in quantum transport at the interface both transmission and reflection occur with a certain probability. The two alternatives can be unified in the single picture of Fig. 4.4. An electron, impinging on the interface from region 1 with wave vector k_1, is

---

[2] Precisely, all possible paths are contributing to the outcome of the process, each with its own phase which is related to the action on that path [65]. In the classical limit, all paths except the classical path interfere destructively, while inside the quantum region some of the non-classical paths interfere constructively, giving rise to observable quantum effects.

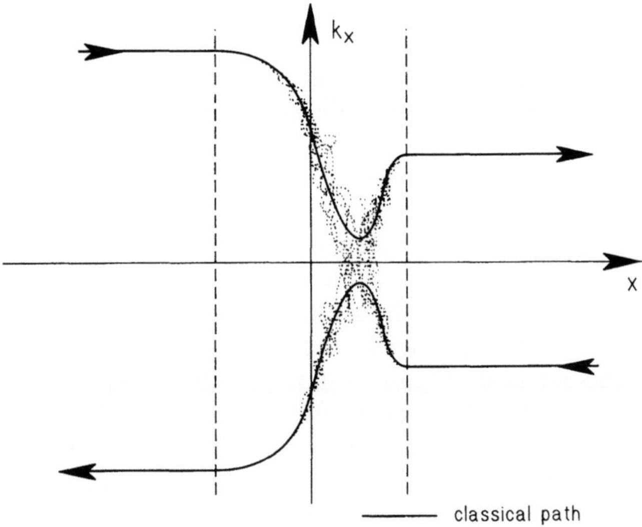

Fig. 4.3 Classical and quantum paths at high electron energies

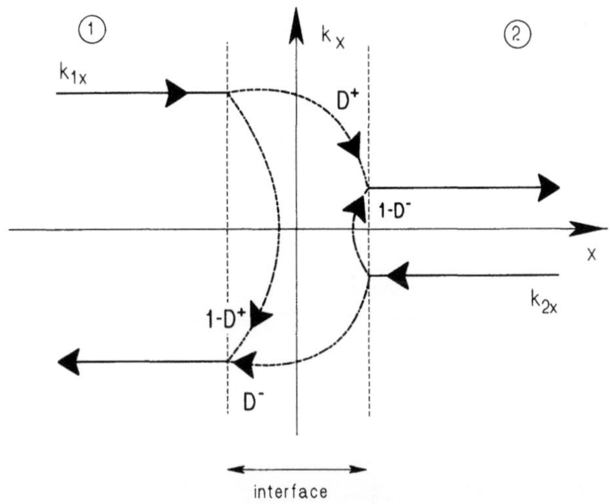

Fig. 4.4 Transmission and reflection at the interface

either transmitted into region 2 with a probability $D^+(\mathbf{k}_1)$ or reflected with probability $1 - D^+(\mathbf{k}_1)$. Likewise, an electron coming from region 2 is transmitted into region 1 with probability $D^-(\mathbf{k}_2)$ or reflected with probability $1 - D^-(\mathbf{k}_2)$.

Thus, with respect to electron transport, the interface is fully characterized by specifying the transmission probability.

This statement completes the characterization of the interface as given in

Chapter 3. In that chapter we concentrated on the electrostatic characterization of the interface in order to arrive at interface conditions for the Poisson equation while disregarding any carrier transport effects. This gave rise to the parameters "band edge discontinuity ΔW_c" and "surface charge density σ", which entered the interface conditions for the Poisson equation. The transmission probability $D(\mathbf{k})$ describes the effect of the interface with regard to carrier transport. It will enter the interface conditions for the Boltzmann equation in the subsequent derivation.

As mentioned previously, the main goal of this book is a macroscopic description of the interface in the sense that the effect of the interface on the transport in the bulk of the device is of interest. Hence, the detailed situation *inside* the interface region is not primarily important. As already discussed in the case of the electrostatic boundary conditions in Chapter 3, it suffices to introduce characteristic quantities that describe the effects of the interface. With this goal in mind, we contract the finite thickness of the interface region into a two-dimensional surface. In this picture, the two bulk regions are separated by an infinitesimally thin interface with interface conditions connecting the solutions of the bulk transport equations. The processes taking place inside the interface region become part of the interface conditions.

This situation is exemplarily depicted in Fig. 4.5. In the upper part of the figure, a possible form of the band edge inside the interface layer is shown in detail. The form of the bands inside the interface determine the trans-

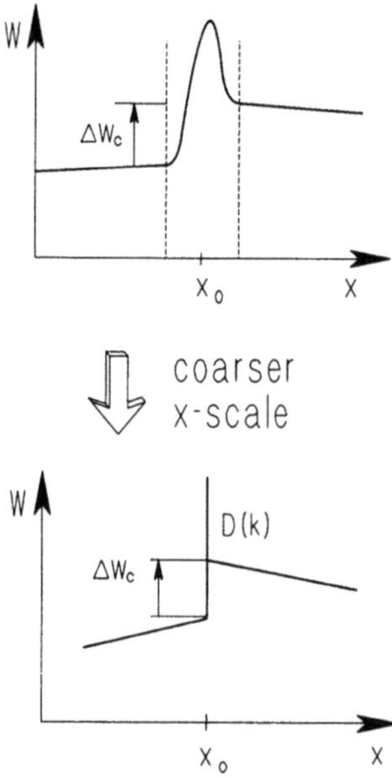

Fig. 4.5 Contraction of the interface region into a surface

mission probability $D(\mathbf{k})$. When considering the function of the device containing the interface, we look at the interface on a coarser scale in position space. The lower part of Fig. 4.5 shows the respective view of the interface. In this macroscopic view, the interface is characterized by ΔW_c, σ, and $D(\mathbf{k})$. These parameters can be obtained from a detailed consideration of the microscopic view of the interface. From the macroscopic point of view, they are assumed as given parameters or models.

Now the ideas of the preceding paragraphs have to be cast into a mathematical formulation. The following assumptions will be made in the derivation:

1. no quantum effects outside the interface region;
2. energy conservation;
3. conservation of transverse crystal momentum;
4. quantum effects inside the interface (reflection, tunneling) are described by $D(\mathbf{k})$.

The first three items comprise full semi-classical transport, while the 4th item adds quantum effects, but only inside the interface. This is what we have termed "semi-quantum" transport above.

Assumption 2, energy conservation, is valid on a not too small time scale [66], thus dynamic quantum effects at the interface have to be excluded. We will assume that the electronic behaviour inside the interface region is instantaneous; delay effects inside the interface are not considered. Respective investigations [67] showed tunneling times in the order of 10^{-15} s for typical barriers of 1 nm thickness, which is much faster than the time scales of the limiting effects in usual devices.

It is shown in Appendix B that conservation of transverse crystal momentum holds if the potential energy does not depend on the tangential direction. Hence, assumption 3 seems to be reasonable if the tangential dependence of the interface potential is weak. Moreover, the conservation conditions 2 and 3 only hold if scattering inside the interface can be neglected, which is possible if the interface region is thin [51]. An extension of the model that includes scattering will be considered in Sections 4.7 and 4.8.

4.2.2 The Relation of the Wave Vectors

The conservation conditions determine a relationship between the wave vectors $\mathbf{k}_1$ and $\mathbf{k}_2$ in the two regions. For simplicity, we begin with the case of a single conduction band in each of the two regions, and with a spherical parabolic energy dispersion law. General $W(\mathbf{k})$-relationships will be considered in Section 4.6, and multiple bands in Section 4.8.

The basic configuration considered is depicted in Fig. 4.6. At the interface, two different materials 1 and 2 adjoin each other. The interface is character-

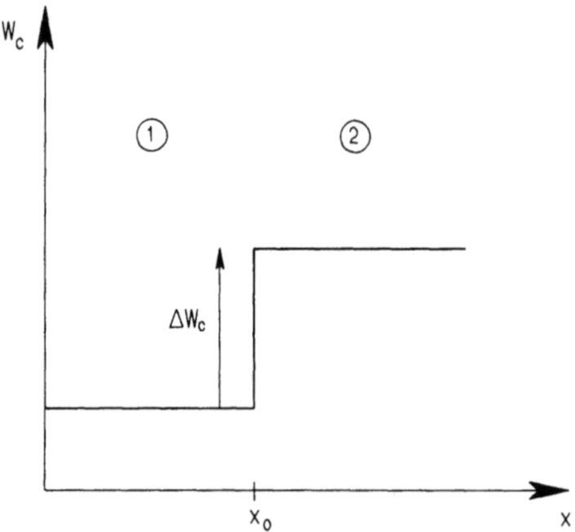

Fig. 4.6 Conduction band discontinuity

ized by a band edge discontinuity ΔW_c. A possible gradient of the band edge at the interface is not visible on the scale of Fig. 4.6. The geometry is chosen such that the interface occurs at the location $\mathbf{x}_0$. The x-axis is pointing in the direction normal to the interface, while the y- and z-axes are the tangential directions. Particles moving in positive x-direction are defined to undergo a positive increase ΔW_c of potential energy while crossing the interface. The case of negative ΔW_c or other orientations of the interface must be treated by appropriate transformations.

According to Fig. 4.4, the two cases transition and reflection have to be distinguished. The first case considered is *transition*. The energy conservation law states equality of the particle's energy on both sides of the interface, with the shift of band structures ΔW_c taken into account

$$W_1(\mathbf{k}_1) = \Delta W_c + W_2(\mathbf{k}_2), \tag{4.5}$$

where $W_i(\mathbf{k}_i)$ represents the band structure of the respective material. Conservation of the tangential crystal momentum reads

$$\hbar k_{1y} = \hbar k_{2y},$$
$$\hbar k_{1z} = \hbar k_{2z}. \tag{4.6}$$

In principle, Eqs. (4.5)–(4.6) are three equations for the 3 components of $\mathbf{k}_2$ if $\mathbf{k}_1$ is known, or vice versa.

In the case of spherical parabolic bands, the energy dispersion is

$$W(\mathbf{k}) = \frac{\hbar^2}{2m}(k_x^2 + k_y^2 + k_z^2) \tag{4.7}$$

with m being the effective mass of the band. Inserting (4.7) into (4.5) yields

$$\frac{\hbar^2}{2m_1}\left[(k_{1x}^2 + k_{1y}^2 + k_{1z}^2) - \frac{m_1}{m_2}(k_{2x}^2 + k_{2y}^2 + k_{2z}^2) - \frac{2m_1}{\hbar^2}\Delta W_c\right]$$
$$= 0. \tag{4.8}$$

Eq. (4.8) with (4.6) can be solved for k_{2x} as a function of $\mathbf{k}_1$, giving

$$k_{2x}(\mathbf{k}_1) = \mathrm{sgn}(k_{1x})\sqrt{\frac{m_2}{m_1}(k_{1x}^2 - \tilde{k}^2)}, \tag{4.9}$$

where the abbreviation

$$\tilde{k} = \sqrt{\frac{2m_1}{\hbar^2}\Delta W_c + \frac{m_1 - m_2}{m_2}(k_y^2 + k_z^2)} \tag{4.10}$$

has been introduced. The index of k_y resp. k_z has been dropped because of
(4.6). We restrict the analysis here to $m_1 > m_2$; thus $\tilde{k}$ is guaranteed to be
real and positive. The case $m_1 < m_2$ will be considered in Section 4.6.
The inverse solution of (4.8) for k_{1x} as a function of $\mathbf{k}_2$ reads

$$k_{1x}(\mathbf{k}_2) = \mathrm{sgn}(k_{2x})\sqrt{\frac{m_1}{m_2}k_{2x}^2 + \tilde{k}^2}. \tag{4.11}$$

From (4.9) it can be seen that the $\mathbf{k}_2$-$\mathbf{k}_1$-relationship has a real solution only
if the condition

$$k_{1x}^2 > \tilde{k}^2 \tag{4.12}$$

holds. This condition reflects the fact that particles impinging on the interface from region 1 must have a sufficiently large normal momentum to be
able to overcome the energy barrier. For the reverse direction, (4.12) expresses the minimum wave vector a particle has after transition from region
2. Thus (4.12) represents the condition for transmission across the interface.
The set of $\mathbf{k}_1$-vectors that satisfies condition (4.12) is depicted by the
hatched area in Fig. 4.7, where k_t is one of k_y or k_z, and $k^* = \sqrt{2m_1\Delta W_c/\hbar^2}$.
For the sake of notation simplicity, we define for the following that the
transmission probability be zero if (4.12) is not satisfied.
On the other side of the interface, a corresponding restriction of the expression under the square root in (4.11) does not exist. Hence all states in
$\mathbf{k}_2$-space—or to put it similar to (4.12), the states satisfying

$$k_{2x}^2 > 0 \tag{4.13}$$

—are able to transit to the other side (or may have come from the other
side if their velocity points away from the interface into region 2).
Now we turn to the case of *reflection*. Particles not satisfying the transmission condition must be reflected at the interface. The following equations
consider the case that a carrier impinges on the interface from region 1 with

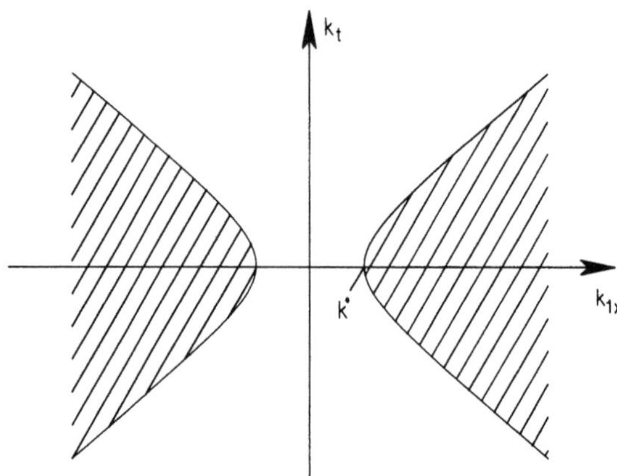

Fig. 4.7 Transition states in $\mathbf{k}_1$-space

wave vector $\mathbf{k}'_1$ and is reflected into region 1 with wave vector $\mathbf{k}_1$. Analogous expressions hold for the other side of the interface.

Conservation of tangential crystal momentum in this case is expressed by

$$\hbar k_{1y} = \hbar k'_{1y},$$
$$\hbar k_{1z} = \hbar k'_{1z}, \tag{4.14}$$

in analogy to (4.6). Since the reflected carrier stays within the same band, and the energy dispersion (4.7) is symmetric with respect to sign reversal of each of the $\mathbf{k}$ components, the conservation of energy in the case of reflection becomes simply

$$k_{1x} = -k'_{1x}. \tag{4.15}$$

In contrast to transmission, reflection is always possible. The fraction of particles that is reflected or transmitted is determined by the actual reflection and transmission probabilities, respectively.

4.2.3 The Interface Condition

Now we are prepared to set up the interface condition for the Boltzmann equation. In consistence with the discussion of the Boltzmann equation in the bulk in Chapter 2, we consider particle balance at the interface. Hence, the interface condition will be based on the following analysis of particle numbers arriving at or leaving from the interface. As discussed in Section 2.5, it is the *inflow* of particles into the respective region that determines the interface condition.

Fig. 4.8 shows the volume elements in phase space containing the particles

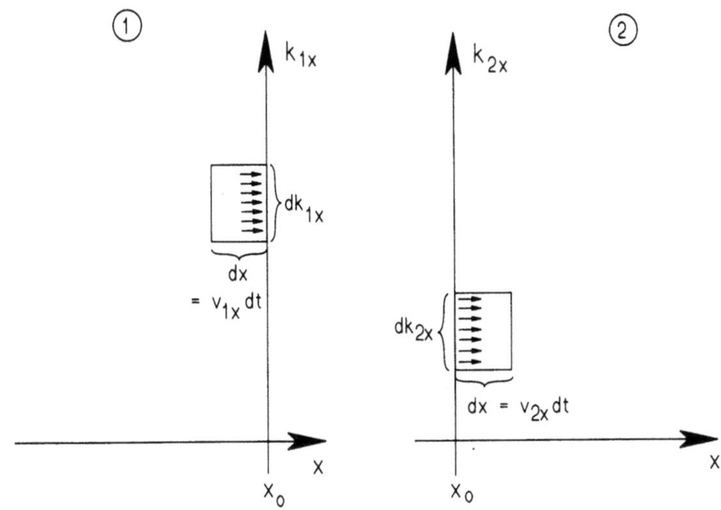

Fig. 4.8 Particle numbers at the interface

that arrive at resp. leave the interface during a time interval dt. The volume elements are taken infinitesimally small, so that all quantities in the elements can be considered as constant. According to Fig. 4.8, the number of particles arriving at the interface from region 1 during a time step dt with wave vector $\mathbf{k}_1$ is proportional to $v_{1x}(\mathbf{k}_1)f_1(\mathbf{x}_0, \mathbf{k}_1)\, d^3\mathbf{k}_1\, dy\, dz\, dt$ [45], while the number of particles leaving the interface with wave vector $\mathbf{k}_2$ into material 2 is proportional to $v_{2x}(\mathbf{k}_2)f_2(\mathbf{x}_0, \mathbf{k}_2))\, d^3\mathbf{k}_2\, dy\, dz\, dt$. The fluxes of particles incident on the interface from region 2 and leaving the interface into region 1 are formed analogously.

As mentioned in Subsection 4.2.1, tunneling effects at the interface are considered by the transmission probability which describes the fraction of the incident flux that crosses the interface. Recall that $D^+(\mathbf{k}_1)$ is the probability that an electron impinging on the interface with wave vector $\mathbf{k}_1$ from region 1 (the positive x-direction) will cross the interface (cf. Fig. 4.4), and $D^-(\mathbf{k}_2)$ the probability of the reverse process. D^+ and D^- are not independent, but transform into each other as $D^+(\mathbf{k}_1) = D^-(\mathbf{k}_2(\mathbf{k}_1))$ because of the time reversal symmetry of the Schrödinger equation. Multiplication of the incident flux by the transmission probability yields the flux actually crossing the interface, while the rest of the flux is reflected.

Now the fluxes into the regions are composed according to Fig. 4.4 by the respective transmitted and reflected particles. The number of particles travelling into material 2 with wave vector $\mathbf{k}_2$ (inflow) consists of the fraction of particles incident from region 1 that cross the interface (transflow) plus the fraction of particles impinging on the interface from region 2 but being reflected (reflow). Transflow and reflow are determined from the number of particles arriving at the interface (outflow) according to the transition and reflection probabilities. Schematically the inflow into region 2 is

$$\text{inflow}(2) = \text{transflow}(1) \quad + \text{reflow}(2)$$
$$= D^+ \cdot \text{outflow}(1) + (1 - D^-) \cdot \text{outflow}(2).$$

The inflow into region 1 is formed analogously.

Inserting the respective particle fluxes, the expressions for the inflow become

$$
\begin{aligned}
v_{2x} f_2(\mathbf{k}_2)\, d^3\mathbf{k}_2 &= v_{1x} D^+(\mathbf{k}_1(\mathbf{k}_2)) f_1(\mathbf{k}_1(\mathbf{k}_2))\, d^3\mathbf{k}_1 \\
&\quad + v_{2x}[1 - D^-(-k_{2x}, k_{2y}, k_{2z})] \\
&\quad \times f_2(-k_{2x}, k_{2y}, k_{2z})\, d^3\mathbf{k}_2 \\
&\quad (v_{2x} > 0), \\
v_{1x} f_1(\mathbf{k}_1)\, d^3\mathbf{k}_1 &= v_{2x} D^-(\mathbf{k}_2(\mathbf{k}_1)) f_2(\mathbf{k}_2(\mathbf{k}_1))\, d^3\mathbf{k}_2 \\
&\quad + v_{1x}[1 - D^+(-k_{1x}, k_{1y}, k_{1z})] \\
&\quad \times f_1(-k_{1x}, k_{1y}, k_{1z})\, d^3\mathbf{k}_1 \\
&\quad (v_{1x} \le 0).
\end{aligned}
$$

$$(4.16)$$
$$(4.17)$$

Eq. (4.16) is the inflow into region 2, while (4.17) is the corresponding equation for inflow into region 1. The relations $\mathbf{k}_2(\mathbf{k}_1)$ resp. $\mathbf{k}_1(\mathbf{k}_2)$ are determined by energy and parallel momentum conservation and the energy band structure, as explained in Subsection 4.2.2.

Eqs. (4.16)–(4.17) constitute the general *interface conditions* for the Boltzmann equation. They consist of two parts, one equation for inflow into each of the two regions separated by the interface. To my knowledge, the conditions have been given in this universality and clarity for the first time in a paper by Maeda [68].

In the case of a boundary, one of the regions separated by the interface is outside the domain of interest. The general *boundary condition* for the Boltzmann equation then consists of that one equation of (4.16)–(4.17) that describes the inflow into the inner region, while the other equation is dropped. The inflow equation for the inner side contains the distribution function on the outer side. Since the distribution function in the environment is not part of the solution, a plausible model for f in the environment must be provided. For instance, in case of a metallic contact, the environment can be considered as a thermal and particle reservoir [69] at a fixed temperature and Fermi level, which can be modelled by the equilibrium distribution.

The notation of (4.16–4.17) still can be somewhat simplified. It is shown in Appendix A by formal variable transformation that the relation

$$v_{2x}(\mathbf{k}_2)\, d^3\mathbf{k}_2 = v_{1x}(\mathbf{k}_1)\, d^3\mathbf{k}_1 \tag{4.18}$$

holds for arbitrary energy band structures. Thus some factors in (4.16)–(4.17) can be cancelled, giving

$$f_2(\mathbf{k}_2) = D^+(\mathbf{k}_1(\mathbf{k}_2))f_1(\mathbf{k}_1(\mathbf{k}_2))$$
$$+ \left[1 - D^-(-k_{2x}, k_{2y}, k_{2z})\right]f_2(-k_{2x}, k_{2y}, k_{2z})$$
$$(v_{2x} > 0), \tag{4.19}$$

$$f_1(\mathbf{k}_1) = D^-(\mathbf{k}_2(\mathbf{k}_1))f_2(\mathbf{k}_2(\mathbf{k}_1))$$
$$+ \left[1 - D^+(-k_{1x}, k_{1y}, k_{1z})\right]f_1(-k_{1x}, k_{1y}, k_{1z})$$
$$(v_{1x} \le 0). \tag{4.20}$$

Although this notation is simpler than (4.16)–(4.17), the nature of a flux balance is no longer visible. Conceptually, the formulation (4.16)–(4.17) is thus to be preferred.

4.3 Semi-Classical Interface Transport

In this section, we turn to a simplification of the interface transport described in Section 4.2 by neglecting all quantum effects at the interface. Assumptions 1–3, listed at the end of Subsection 4.2.1, remain in effect; assumption 4 is replaced by the demand that quantum effects must be negligible also at the interface. This means that quantum effects are denied at all, in the bulk as well as in the interface layer. Since the quantum nature of electron transport is now reflected anymore only in the band structure, the model is called "semi-classical transport". As in section 4.2, we restrict the analysis to the simpler case of parabolic bands.

While in semi-quantum transport the transmission probability D could take on any value between 0 and 1, depending on the form of the barrier and the particle energy, in the semi-classical case D is either 0 or 1. This means that the particle is either transmitted or reflected, depending on whether its energy is large enough to surmount the interface barrier or not. This effect is also named *thermionic emission*, which is related to the fact that at higher temperatures more electrons have a sufficient energy for transition.

The coarse-level picture of the energy bands at the interface looks now like Fig. 4.9. Compared to the lower picture of Fig. 4.5, no energy peak at the interface is present which has to be tunnelled through by the carriers. Semi-classically, the interface is a simple step in potential energy.

Hence, if a transmission is possible at all, it takes place with certainty; quantum reflection does not occur. Since the possibility of a transition is determined by the condition (4.12), the transmission probability $D(\mathbf{k})$ becomes

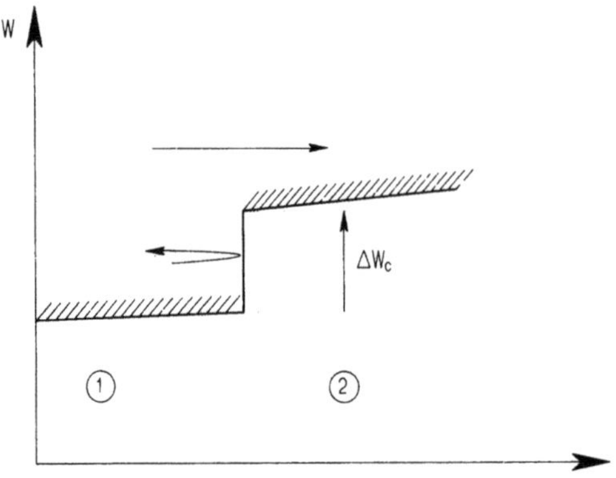

Fig. 4.9 Energy bands at the interface in the semi-classical case

$$D^+(k_{1x}, k_{1y}, k_{1z}) = s(k_{1x} - \tilde{k}) \tag{4.21}$$

for transition in $+x$-direction, and

$$D^-(k_{2x}, k_{2y}, k_{2z}) = 1 \tag{4.22}$$

in the reverse direction, where $s(k)$ is the unit step function (unity for positive k and zero for negative k). Eq. (4.21) expresses the fact that overcoming the potential step is possible only with sufficient energy. Eq. (4.22) says that classically the particle "falls down" the step with certainty.

In order to insert these transmission probabilities into the interface conditions (4.16)–(4.17), D^+ has to be expressed also as a function of $\mathbf{k}_2$ by using the relation $\mathbf{k}_1(\mathbf{k}_2)$, and D^- has to be rewritten as a function of $\mathbf{k}_1$ by $\mathbf{k}_2(\mathbf{k}_1)$. With (4.11), D^+ in (4.21) turns to

$$D^+(\mathbf{k}_1(\mathbf{k}_2)) = 1 \tag{4.23}$$

because $|k_{1x}|$ is always greater than $\tilde{k}$ from (4.11). The relation $\mathbf{k}_2(\mathbf{k}_1)$ (4.9) is only valid under the condition (4.12), thus D^- becomes

$$D^-(\mathbf{k}_2(\mathbf{k}_1)) = s(-k_{1x} - \tilde{k}) \tag{4.24}$$

where $k_{1x} < 0$ for D^- has been taken into account.

With the property of the step function $1 - s(k) = s(-k)$, insertion of Eqs. (4.21)–(4.24) into (4.16)–(4.17) gives the general thermionic emission interface condition for the Boltzmann equation

$$v_{2x} f_2(\mathbf{k}_2)\, d^3\mathbf{k}_2 = v_{1x} f_1(\mathbf{k}_1(\mathbf{k}_2))\, d^3\mathbf{k}_1 \qquad (v_{2x} > 0), \tag{4.25}$$

$$v_{1x} f_1(\mathbf{k}_1)\, d^3\mathbf{k}_1 = v_{2x} f_2(\mathbf{k}_2(\mathbf{k}_1)) \cdot s(-k_{1x} - \tilde{k})\, d^3\mathbf{k}_2$$

$$+ v_{1x} f_1(-k_{1x}, k_{1y}, k_{1z})$$

$$\cdot s(k_{1x} + \tilde{k})\, d^3\mathbf{k}_1 \qquad (v_{1x} \leq 0). \tag{4.26}$$

Note that the probabilities in (4.21)–(4.24) become zero if transition is impossible because of the step functions.

The interface conditions (4.25)–(4.26) are frequently used in device simulations of hetero-devices (see e.g. [70, 71]), where the Boltzmann equation is solved by the Monte Carlo method.

4.4 Interface Conditions for the Balance Equations

While the preceding sections of this chapter dealt with elementary things like transition processes of single carriers—which resulted in the interface conditions for the Boltzmann equation—, this section will introduce interface conditions on a more macroscopic scale. Since device simulation mainly is done on the basis of transport models on a coarser scale than the Boltzmann equation, these conditions will be the foundation for the respective modelling of electron transport in the following chapters.

In Section 2.3 it has been demonstrated how approximative transport models for the semiconductor bulk can be derived from the Boltzmann equation with the method of moments. This section introduces a method for deriving interface conditions that are consistent with the volume transport model. The basic idea is to apply the approximation procedure for the Boltzmann equation in the bulk analogously to the conditions of the Boltzmann equation at the interface. The procedure is thus the following.

1. Take the assumption for $f(\mathbf{x}, \mathbf{k})$ of the transport model;
2. Insert it into the interface condition for the Boltzmann equation;
3. Multiply with $M(\mathbf{k})$;
4. Integrate with respect to $\mathbf{k}$.

The result of this procedure is a system of equations which connect the parameters of the distribution on both sides of the interface. If the number of resulting equations equals the number of unknowns (the parameters), this equation system self-consistently completes the approximate transport model with respective interface conditions.

However, some items have to be regarded compared to the procedure in the volume. The first point is that an assumption of the form of the distribution function which is good in the volume might be bad in the vicinity of the interface. In fact, it can be expected that the presence of the interface produces additional distortions of the distribution function as compared to the volume. If these distortions are of importance, it would be preferrable to use an ansatz that allows for the distortions, and transforms into the bulk distribution when going away from the interface. In many practical cases, however, one is satisfied with a proper description in the interior of the simulation domain, several mean free paths away from the interface [13]. In these cases, the interface conditions resulting from use of the bulk distributions are sufficient. This point will be further discussed in subsequent chapters when we come to the treatment of the specific interface types.

Second, while the Boltzmann equation consists of one equation defined in the total $\mathbf{k}$-space, the interface conditions (4.16)–(4.17) consist of two equations, each defined on only half of $\mathbf{k}$-space. This has two important consequences.

If we compute the moments of (4.16) or (4.17), the integration range must extend only over the inflow portions of $\mathbf{k}_2$ or $\mathbf{k}_1$, respectively. These are the respective domains of validity of Eqs. (4.16) or (4.17). This restriction is consistent with the physical concept that only in the inflow regions the behaviour of the distribution function is influenced from outside.

The second consequence has to do with the choice of $M(\mathbf{k})$. Care has to be taken because of the fact that each moment of the volume equation (2.4) yields *one* first order differential equation, while each moment of the interface condition (4.16)–(4.17) gives *two* equations. So we get two interface conditions for each volume equation, while we need only one (see Section 2.5). Thus one $M(\mathbf{k})$ for the interface condition has to be related to a *pair* of volume equations. In selecting the right $M(\mathbf{k})$ we have to take care that the physical content of (4.16)–(4.17), i.e. the balance of flows, is correctly translated into the picture of the averaged quantities.

In general, the even moments of the Boltzmann equation yield conservation laws of some quantity, .e.g. particle concentration for $M = 1$ or energy density for $M = W(\mathbf{k})$. The corresponding flux quantity is the moment derived from $\mathbf{v}(\mathbf{k})M(\mathbf{k})$. Thus, we find that the *even* moments of (4.16)–(4.17) contain the normal component of the flux $v_x(\mathbf{k})M(\mathbf{k})$. Hence the even $M(\mathbf{k})$ have to be selected for the interface condition, because they yield consistent interface conditions specifying the inflow of the quantity into the respective region. Each even $M(\mathbf{k})$ yields an interface condition for the conservation equation associated with M, as well as an interface condition for the respective flux equation.

Take for instance the drift-diffusion transport model which can be derived from the Boltzmann equation by computing the zero and first order moments (see Section 2.3). Thus, according to Section 2.5, two boundary conditions are necessary to solve this equation system. Setting $M(\mathbf{k})$ to 1 when computing the moments of (4.16)–(4.17) yields two boundary conditions which are related to the particle inflow into the domain under consideration. These boundary conditions are sufficient for the solution of the drift-diffusion equations.

The above discussion suggests the idea that transport equations should always come pairwise, one conservation law and one equation for the flux of the conserved quantity. Indeed, Baccarani and Wordeman found that neglecting the heat flux equation when using energy balance resulted in unphysical effects and numerical instability [72]. Also Duderstadt and Martin recommend to use only an even number of transport equations

Table 4.1. *Interface conditions for the moment equations (Inflow Moments)*

$$\int_0^\infty dk_{2x} \int_{-\infty}^\infty dk_{2y}\, dk_{2z}\, v_{2x}(\mathbf{k}_2) M(\mathbf{k}_2) f_2(\mathbf{k}_2)$$

$$= \int_0^\infty dk_{2x} \int_{-\infty}^\infty dk_{2y}\, dk_{2z}\, v_{2x}(\mathbf{k}_2) M(\mathbf{k}_2)$$

$$\times \{ D^+(\mathbf{k}_1(\mathbf{k}_2)) f_1(\mathbf{k}_1(\mathbf{k}_2))$$

$$+ [1 - D^-(-k_{2x}, k_{2y}, k_{2z})] f_2(-k_{2x}, k_{2y}, k_{2z}) \} \qquad (4.27)$$

$$\int_{-\infty}^0 dk_{1x} \int_{-\infty}^\infty dk_{1y}\, dk_{1z}\, v_{1x}(\mathbf{k}_1) M(\mathbf{k}_1) f_1(\mathbf{k}_1)$$

$$= \int_{-\infty}^0 dk_{1x} \int_{-\infty}^\infty dk_{1y}\, dk_{1z}\, v_{1x}(\mathbf{k}_1) M(\mathbf{k}_1)$$

$$\times \{ D^-(\mathbf{k}_2(\mathbf{k}_1)) f_2(\mathbf{k}_2(\mathbf{k}_1))$$

$$+ [1 - D^+(-k_{1x}, k_{1y}, k_{1z})] f_1(-k_{1x}, k_{1y}, k_{1z}) \} \qquad (4.28)$$

when truncating the infinite equation hierarchy [13] because of otherwise occurring boundary condition problems.

In summary, multiplication of (4.16) and (4.17) with an even function $M(\mathbf{k})$ and integration over the respective inflow region thus yields the general form of the inflow integrals in Table 4.1. These equations constitute the general interface condition for the moment equations. The actual moment is defined by the form of $M(\mathbf{k})$, i.e. zero order moment for $M = 1$ etc. Note that the integration region in any case covers only half of k-space. In contrast to (4.16)–(4.17), the integration variable in (4.27) and in (4.28) has been transformed uniformly to $\mathbf{k}_2$ or $\mathbf{k}_1$, respectively, making use of (4.18).

Since the procedure is in analogy to the method of moments in the volume, but is carried out for the interface with respect to the inflow regions of $\mathbf{k}$-space, it may be called *Inflow Moments Method* [73]. A special case of the formulation, namely its application to boundaries with zero inflow, is known as the Marshak boundary condition [13].

In accordance with the interface conditions for the Boltzmann equation, the moment interface conditions (4.27)–(4.28) consist of two equations: Each of the equations describes the inflow of the respective conservation quantity into one of the regions separated by the interface.

This is further illustrated for the simplified one-dimensional case (with finite Brillouin zone) in Fig. 4.10. The inflow condition (4.16) for region 2 is defined in the upper half of the figure, while the inflow condition into region 1 (4.17) is defined in the lower half. The arrows denote the regions for the respective integrations. While the flux integration over full $\mathbf{k}$-space gives the total flux density, the integration over the indicated regions result

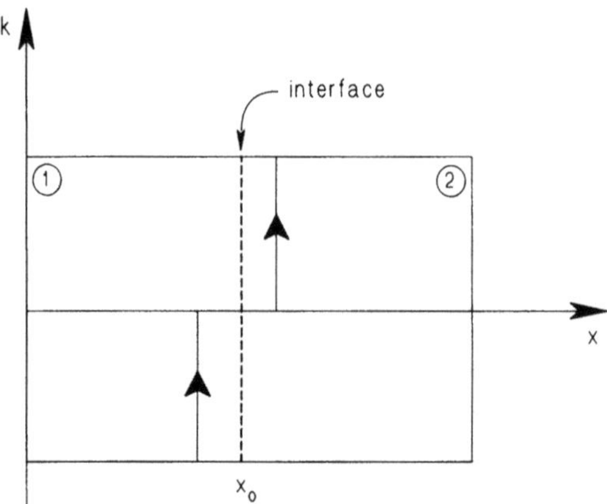

Fig. 4.10 Integration regions for the Inflow Moments Method

in the fluxes corresponding only to the inflowing particles. The sum of the
inflow fluxes will yield a statement on the total flux again.
It is clear that some accuracy is lost in the process of the Inflow Moments
Method, just as we loose "Boltzmann completeness" [74] when going from
the Boltzmann equation to an approximate transport model. In this context, the boundary condition has to be seen as a coarse-level description of
electron transport at the interface on a scale where microscopic features are
not visible. It is only suited for a detailed local description of the distribution at the interface itself if the assumption allows for such detail.

4.5 Semi-Classical Interface Conditions for the Balance Equations

Corresponding to Section 4.3, the interface condition of the previous section can be specialized to the case of semi-classical transport, i.e. thermionic
emission at the interface.
Inserting (4.23) and (4.22) into (4.27) yields the condition

$$
\int_0^\infty dk_{2x} \int_{-\infty}^\infty dk_{2y}\, dk_{2z}\, v_{2x}(\mathbf{k}_2) M(\mathbf{k}_2) f_2(\mathbf{k}_2)
$$
$$
= \int_0^\infty dk_{2x} \int_{-\infty}^\infty dk_{2y}\, dk_{2z}\, v_{2x}(\mathbf{k}_2) M(\mathbf{k}_2) f_1(\mathbf{k}_1(\mathbf{k}_2))
$$

(4.29)

for the inflow into the region on the upper side of the potential step (cf. Eq.
(4.25)).
Eq. (4.28) becomes with (4.24) and (4.21)

$$\int_{-\infty}^{\infty} dk_{1y}\, dk_{1z} \int_{-\infty}^{0} dk_{1x}\, v_{1x}(\mathbf{k}_1) M(\mathbf{k}_1) f_1(\mathbf{k}_1)$$

$$- \int_{-\infty}^{\infty} dk_{1y}\, dk_{1z} \int_{-\tilde{k}}^{0} dk_{1x}\, v_{1x}(\mathbf{k}_1) M(\mathbf{k}_1) f_1(-k_{1x}, k_{1y}, k_{1z})$$

$$= \int_{-\infty}^{\infty} dk_{1y}\, dk_{1z} \int_{-\infty}^{-\tilde{k}} dk_{1x}\, v_{1x}(\mathbf{k}_1) M(\mathbf{k}_1) f_2(\mathbf{k}_2(\mathbf{k}_1)), \qquad (4.30)$$

where the terms containing f_1 have been collected on the left-hand side, and the integration intervals have been truncated according to the step functions in (4.24) and (4.21) (see also Eq. (4.26)). Reversing the sign of k_{1x} according to the variable transformation

$$\int_{-\tilde{k}}^{0} dk_{1x}\, v_{1x}(\mathbf{k}_1) M(\mathbf{k}_1) f_1(-k_{1x}, k_{1y}, k_{1z})$$

$$= \int_{0}^{\tilde{k}} dk_{1x}\, v_{1x}(-k_{1x}, k_{1y}, k_{1z}) M(-k_x, k_{1y}, k_{1z}) f_1(\mathbf{k}_1), \qquad (4.31)$$

Eq. (4.30) becomes

$$\int_{-\infty}^{\infty} dk_{1y}\, dk_{1z} \left[\int_{-\infty}^{0} dk_{1x}\, v_{1x}(\mathbf{k}_1) M(\mathbf{k}_1) f_1(\mathbf{k}_1) \right.$$

$$\left. - \int_{0}^{\tilde{k}} dk_{1x}\, v_{1x}(-k_{1x}, k_{1y}, k_{1z}) M(-k_x, k_{1y}, k_{1z}) f_1(\mathbf{k}_1) \right]$$

$$= \int_{-\infty}^{\infty} dk_{1y}\, dk_{1z} \int_{-\infty}^{-\tilde{k}} dk_{1x}\, v_{1x}(\mathbf{k}_1) M(\mathbf{k}_1) f_2(\mathbf{k}_2(\mathbf{k}_1)). \qquad (4.32)$$

This equation can be simplified further by recalling that $M(\mathbf{k})$ is an even function:

$$M(\mathbf{k}_1) = M(-k_{1x}, k_{1y}, k_{1z}). \qquad (4.33)$$

In addition, the velocity (2.5) is an odd function of $\mathbf{k}$:

$$v_x(\mathbf{k}_1) = -v_x(-k_{1x}, k_{1y}, k_{1z}). \qquad (4.34)$$

With (4.33) and (4.34) the terms on the left-hand side of (4.32) can be combined, giving

$$\boxed{\begin{aligned} \int_{-\infty}^{\infty} dk_{1y}\, dk_{1z} \int_{-\infty}^{+\tilde{k}} dk_{1x}\, v_{1x}(\mathbf{k}_1) M(\mathbf{k}_1) f_1(\mathbf{k}_1) \\ = \int_{-\infty}^{\infty} dk_{1y}\, dk_{1z} \int_{-\infty}^{-\tilde{k}} dk_{1x}\, v_{1x}(\mathbf{k}_1) M(\mathbf{k}_1) f_2(\mathbf{k}_2(\mathbf{k}_1)) \end{aligned}} \qquad (4.35)$$

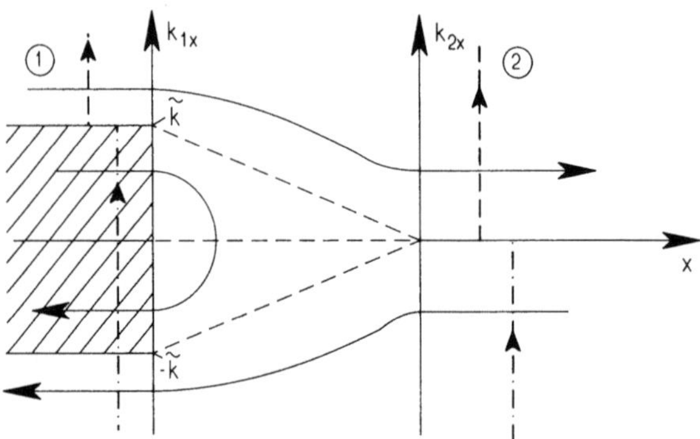

Fig. 4.11 Flux balance in thermionic emission

as the final result for the condition of inflow into region 1 in the thermionic emission case.

The fluxes appearing in Eqs. (4.29) and (4.35) can be visually interpreted in Fig. 4.11. All carriers arriving at the interface with $k_{1x} < \tilde{k}$ (the hatched area) are reflected to $-k_{1x}$, while all particles arriving with $k_{1x} > \tilde{k}$ cross the interface, being slowed down. In the reverse direction, all electrons arriving at the interface transit to region 1 and are accelerated, having a longitudinal wave vector k_{1x} with an absolute value of at least $\tilde{k}$.

For interpretational purpose, (4.29) and (4.35) are rewritten such that the integration variable corresponds to the region where the respective distribution function belongs to:

$$\int_0^\infty dk_{2x} \int_{-\infty}^\infty dk_{2y}\, dk_{2z}\, v_{2x}(\mathbf{k}_2) M(\mathbf{k}_2) f_2(\mathbf{k}_2)$$
$$= \int_{-\infty}^\infty dk_{1y}\, dk_{1z} \int_{+\tilde{k}}^\infty dk_{1x}\, v_{1x}(\mathbf{k}_1) M(\mathbf{k}_2(\mathbf{k}_1)) f_1(\mathbf{k}_1) \tag{4.36}$$

$$\int_{-\infty}^\infty dk_{1y}\, dk_{1z} \int_{-\infty}^{+\tilde{k}} dk_{1x}\, v_{1x}(\mathbf{k}_1) M(\mathbf{k}_1) f_1(\mathbf{k}_1)$$
$$= \int_{-\infty}^\infty dk_{2y}\, dk_{2z} \int_{-\infty}^0 dk_{2x}\, v_{2x}(\mathbf{k}_2) M(\mathbf{k}_1(\mathbf{k}_2)) f_2(\mathbf{k}_2). \tag{4.37}$$

The integration intervals of (4.36) and (4.37) are depicted in Fig. 4.11 by the thick dashed and dash-dotted lines, respectively. Comparing (4.36) and Fig. 4.11 it can be seen that the total inflow into region 2 comes from all carriers in region 1 with $k_{1x} > \tilde{k}$. From (4.37) follows that the total outflow from region 2 equals the flow integral in region 1 with respect to the interval from $-\infty$ to $+\tilde{k}$, where the portion from $-\tilde{k}$ to $+\tilde{k}$ consistently covers the reflected carriers. After exchanging the left-hand side and the right-hand side of (4.36) and adding both equations, the integration intervals can be combined to the full space and thus give the relation of the total fluxes on both sides of the interface.

4.6 General Band Structures

While the preceding sections introduced the fundamental relationships on the basis of parabolic bands, this section examines the case of general energy dispersions laws. Still, we consider only a single band on each side of the interface. Multiple bands are treated in Section 4.8.

First we determine the general k_1-k_2-relationship from the conservation laws. This problem involves the inversion of the band functions. Thus we define the inverse K of the energy dispersion $W(\mathbf{k})$ by

$$W(k_x, \mathbf{k}_t) = \overline{W} \quad \Leftrightarrow \quad k_x = K(\overline{W}, \mathbf{k}_t), \tag{4.38}$$

where $\mathbf{k}_t$ is the transverse part of the vector $\mathbf{k}$: $\mathbf{k} = (k_x, k_y, k_z) = (k_x, \mathbf{k}_t)$, and $\overline{W}$ is some (fixed) energy. The arguments of K can be combined to a vector of a three-dimensional vector space consisting of an energy and the two components of $\mathbf{k}_t$; we shall briefly call it the WKT-space.

In general, there are multiple solutions of (4.38) for a given $\overline{W}$. To be precise in this point, the result of K is defined to be the total set of k_x which satisfies (4.38). Only two solutions are possible if the surface of constant energy in $\mathbf{k}$-space does not have any recessed portions. (Mathematically, the surface points must be of the same type everywhere, i.e. either elliptic or hyperbolic [75].) Otherwise, more than two solutions may occur. We will consider in the following only the simpler case of two possible solutions.

The function $W(\mathbf{k})$ is only invertible, i.e. the result set of K is not empty, if the energy in the argument of K lies in the energy interval of the band $W(\mathbf{k})$ at all. This means that the full argument vector of K must be in a set V of vectors in WKT-space that is formally defined by

$$V = \left\{ (W, \mathbf{k}_t) : \bigvee_{k_x} W = W(k_x, \mathbf{k}_t) \right\}. \tag{4.39}$$

The notation means: "V is the set of all vectors $(\overline{W}, \mathbf{k}_t)$ for which holds: A component k_x exists such that $\overline{W}$ equals $W(k_x, \mathbf{k}_t)$".

4.6.1 Transition

In the case of an interface, W, K and V are defined for each of the two materials separated by the interface; the material will be indicated by the respective index. The system of bands on each side have a band offset of ΔW_c; the energy conservation of electrons transiting between any of the bands is thus given by (4.5), where W_1 and W_2 are the energies of the respective bands under consideration (band indices suppressed). In this general formulation, however, the assumption $\Delta W_c > 0$ is no longer necessary.

Since an electron arriving at the interface from region 1 is of course in a valid state of the energy band W_1, it occupies a point in V_1, while an electron leaving into region 2 represents a point in V_2. A transition from 1 to 2 (or vice versa) is only possible if Eqs. (4.5)–(4.6) can be simultaneously satisfied. In the language of the WKT-space, this means that the point in V_1, after shifting the energy component by the band offset ΔW_c, must also be a valid point in V_2 (cf. also the "overlap of the projection surfaces" discussed in [76]). Thus it is convenient to define the shifted V_1 resp. V_2 by

$$V_1^* = \{(W_2, \mathbf{k}_t): (W_2 + \Delta W_c, \mathbf{k}_t) \in V_1\}, \tag{4.40}$$

$$V_2^* = \{(W_1, \mathbf{k}_t): (W_1 - \Delta W_c, \mathbf{k}_t) \in V_2\}. \tag{4.41}$$

For the moment, W_1 and W_2 denote arbitrary values of energy. With this notation, we can simply say that a transition is possible if the point in V_1 is also a point in V_2^* (whether the transition goes from 1 to 2 or vice versa does not play a role in this condition). In other words, if a point in WKT-space belongs at the same time to V_1 and to V_2^*, then Eqs. (4.5)–(4.6) can be simultaneously satisfied. Equivalently we can say that the point must be in V_2 *and* V_1^* at the same time.

The set of all points belonging to V_1 as well as to V_2^* is given by the intersection Z_1 of the two sets:

$$Z_1 = V_1 \cap V_2^*. \tag{4.42}$$

Analogously, we define the intersection of V_2 and V_1^* by

$$Z_2 = V_2 \cap V_1^*. \tag{4.43}$$

Z_1 and Z_2 can be seen as a generalization of the "overlap region of the shadows onto the y-z-plane of the energy surfaces of the two regions", which has been introduced by Fonash [77] (see also [76]) in order to obtain a formulation for the $\mathbf{k}$-vectors that can participate in a transition. In fact, Fonash' overlap region is recovered by taking a cross section of Z_1 or Z_2 at a particular energy.

Now we can introduce the set T_1 of all $\mathbf{k}_1$ for which a transition is possible by

$$T_1 = \{\mathbf{k}_1: (W_1(\mathbf{k}_1), k_{1y}, k_{1z}) \in Z_1\}, \tag{4.44}$$

saying that an electron in region 1 is able to transit if it is in Z_1. Accordingly, the set T_2 of all $\mathbf{k}_2$ for which a transition is possible becomes

$$T_2 = \{\mathbf{k}_2: (W_2(\mathbf{k}_2), k_{2y}, k_{2z}) \in Z_2\}. \tag{4.45}$$

T_1 and T_2 include both directions of transition. The sets can be decomposed according to

$$T_n = T_n^- + T_n^+ \qquad (n = 1, 2), \tag{4.46}$$

where T_n^+ and T_n^- contain the states with positive and negative velocity, respectively.

The sets for which transitions are *not* possible are defined analogously:

$$B_1 = \{\mathbf{k}_1 : \mathbf{k}_1 \notin T_1\}, \tag{4.47}$$

$$B_2 = \{\mathbf{k}_2 : \mathbf{k}_2 \notin T_2\}. \tag{4.48}$$

The sets can also be decomposed further similar to (4.46).

The generalization of the condition for transition (4.12) becomes now

$$\mathbf{k}_1 \in T_1, \tag{4.49}$$

while the generalization of (4.13) becomes

$$\mathbf{k}_2 \in T_2. \tag{4.50}$$

From (4.50) we can infer that in contrast to (4.13) it may not be possible in general for every electron on the top side of the potential step to cross the interface. This depends on the detailed situation, since not every state approaching the interface may be an element of T_2. The example below will show such a case.

Finally, the relation between $\mathbf{k}_2$ and $\mathbf{k}_1$ in the general formulation is

$$\mathbf{k}_1(\mathbf{k}_2) = (K_1(W_2(\mathbf{k}_2) + \Delta W_c, k_{2y}, k_{2z}), k_{2y}, k_{2z}) \qquad (\mathbf{k}_2 \in T_2), \tag{4.51}$$

while the inverse relation reads

$$\mathbf{k}_2(\mathbf{k}_1) = (K_2(W_1(\mathbf{k}_1) - \Delta W_c, k_{1y}, k_{1z}), k_{1y}, k_{1z}) \qquad (\mathbf{k}_1 \in T_1). \tag{4.52}$$

The conditions behind the equations take care that the energy and transverse momentum conditions can be satisfied. From the two results of K that one has to be selected such that $v_{1x}(\mathbf{k}_1)$ and $v_{2x}(\mathbf{k}_2)$ have the same sign.

4.6.2 Reflection

The case of reflection at arbitrary band structures is quite similar to the transition case. The difference is that now only one band is involved, and that the velocity changes its sign during the process. The general principle is the same, i.e. conservation of energy and transverse crystal momentum. When an electron is reflected at the interface, the transverse components k_y

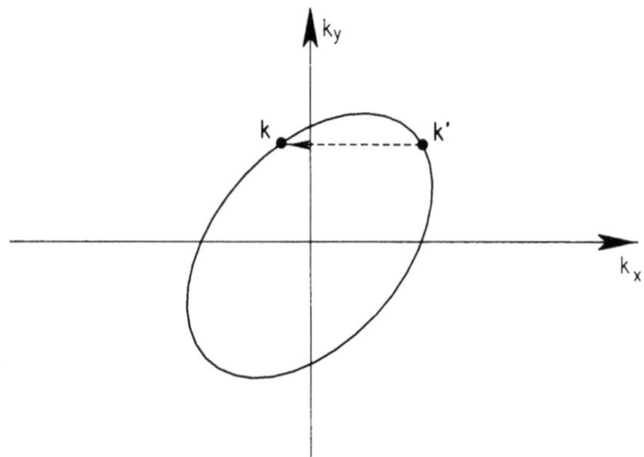

Fig. 4.12 Change of wave vector components at reflection

and k_z of the wave vector are preserved, while the normal component adjusts such that the energy is conserved *and* the normal component of the velocity points away from the interface. The situation is depicted in Fig. 4.12, where an assumed energy surface of some band is shown together with an example of a pair of incident and reflected states. Because of transverse momentum conservation, the reflected state lies on a line parallel to the k_x-axis going through the incident state. Because of energy conservation, the reflected state is determined by the crossing point of this line with the energy surface.

While the aforementioned conditions have to be satisfied if a transition can take place, it follows from the periodicity of the band structure in the repeated zone scheme [25] that reflection is always possible.

Let us denote the incident state by $\mathbf{k}'$ and the reflected one by $\mathbf{k}$; the band structure is given as $W(\mathbf{k})$. Energy conservation reads in this case

$$W(\mathbf{k}) = W(\mathbf{k}'), \tag{4.53}$$

while transverse momentum conservation becomes

$$\hbar k_y = \hbar k_y',$$
$$\hbar k_z = \hbar k_z'. \tag{4.54}$$

From (4.53) and (4.54) follows with (4.38) that the relation between $\mathbf{k}$ and $\mathbf{k}'$ is (see also Appendix A)

$$\mathbf{k}(\mathbf{k}') = \begin{bmatrix} k_x(\mathbf{k}') \\ k_y(\mathbf{k}') \\ k_z(\mathbf{k}') \end{bmatrix} = \begin{bmatrix} K(W(\mathbf{k}'), k_y', k_z') \\ k_y' \\ k_z' \end{bmatrix}, \qquad (v_x(\mathbf{k}) = -v_x(\mathbf{k}')). \tag{4.55}$$

According to the definition of K, the result of K contains both the incident state itself and the reflected state (cf. Fig. 4.12). The condition in (4.55) selects from these the reflected state.

4.6.3 An Example

As an instructive example, we apply the formalism to parabolic bands again, but now drop the condition $m_2 < m_1$. Fig. 4.13 shows the energy dispersion of the two bands. Note from the curvature in the figure that the case $m_2 > m_1$ is shown. The range of the k-vectors is assumed to be unrestricted.

Taking (4.7) for band 1 and solving for k_{1x} gives with (4.38) the inverse energy dispersion

$$K_1(W_1, \mathbf{k}_t) = \pm \sqrt{\frac{2m_1}{\hbar^2} W_1 - k_t^2}. \tag{4.56}$$

Note that the square root only has a real solution if the argument of K_1 is in the set V_1, defined by

$$V_1 = \left\{ (W_1, \mathbf{k}_t) \colon W_1 \geq \frac{\hbar^2}{2m_1} k_t^2 \right\}. \tag{4.57}$$

Only if the condition in (4.57) is met, a portion of W_1 is left which can be carried by k_{1x}. Otherwise, already the transverse wave vector carries an energy higher than W_1, leaving no energy for k_{1x} to contribute. Analogously, the corresponding set V_2 for band 2 becomes

$$V_2 = \left\{ (W_2, \mathbf{k}_t) \colon W_2 \geq \frac{\hbar^2}{2m_2} k_t^2 \right\}. \tag{4.58}$$

Again, we assume that band 2 is shifted by the amount ΔW_c in relation to band 1. To be specific, we take again $\Delta W_c > 0$; otherwise, as we will see in

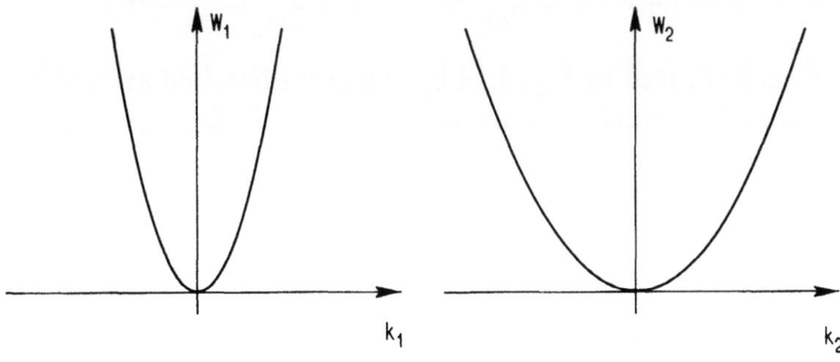

Fig. 4.13 Two parabolic bands

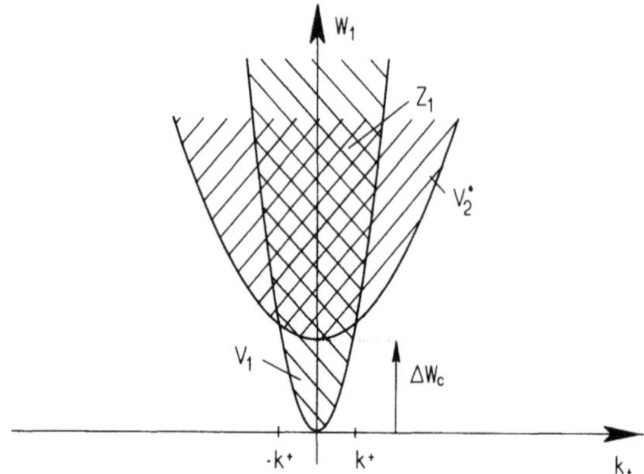

Fig. 4.14 Sets of permissible states

a moment, all equations to follow remain the same with indices 1 and 2 interchanged.

In order to compare the energies in V_1 and V_2, we shift W_2 in (4.58) according to (4.41) by ΔW_c, arriving at

$$V_2^* = \left\{ (W_1, \mathbf{k}_t): W_1 \geq \frac{\hbar^2}{2m_2} k_t^2 + \Delta W_c \right\}. \tag{4.59}$$

The energies in V_1 and V_2^* are now counted both with respect to the origin of band 1 and can thus be compared (for this reason the energy notation in V_2^* has been changed to W_1). The sets V_1 and V_2^* are graphically displayed in Fig. 4.14. The hatched areas in Fig. 4.14 indicate the sets V_1 and V_2^*; according to (4.57) and (4.59) only the points lying "inside" the parabolas belong to the sets.

Transitions are only possible for states belonging both to V_1 and V_2^*, i.e. states in the intersection Z_1 of the two sets. For Z_1, the conditions in both (4.57) and (4.59) must be satisfied in common, thus

$$Z_1 = \left\{ (W_1, \mathbf{k}_t): W_1 \geq \frac{\hbar^2}{2m_1} k_t^2 \wedge W_1 \geq \frac{\hbar^2}{2m_2} k_t^2 + \Delta W_c \right\}. \tag{4.60}$$

The set Z_1 is indicated in Fig. 4.14 by the cross-hatched area; it is the area where V_1 and V_2^* overlap. Only for states in Z_1 Eqs. (4.5)–(4.6) can be satisfied.

Looking at Fig. 4.14, we can observe a remarkable feature. If, in contrast to the picture, m_1 exceeds m_2, the boundary of V_2^* is curved stronger than that of V_1, and the boundaries never intersect. Likewise, the boundaries do not cross in Fig. 4.14 if ΔW_c is negative. In these cases, Z_1 is identical to V_2^*, and electrons in all of $\mathbf{k}_2$-space are able to transit to region 1, as has been found in Subsection 4.2.2, Eq. (4.13). On the other hand, if $m_1 < m_2$ there are points in V_1 as well as in V_2^* for which a transition is forbidden. This means that region 1 and also region 2 have states for which transitions are not allowed. If we simultaneously change the sign of ΔW_2 and make m_2 smaller than m_1, Fig. 4.14 remains qualitatively unaltered.

A formal representation of Z_1 is obtained from (4.60) by evaluating the bounds of the conditions and determining the crossover points. The result is

$$
Z_1 = \left\{ (W_1, \mathbf{k}_t): \begin{cases} W_1 \geq \dfrac{\hbar^2}{2m_1} k_t^2 & \text{if } k_t^2 \geq k^{+2} \\[2ex] W_1 \geq \dfrac{\hbar^2}{2m_2} k_t^2 + \Delta W_c & \text{if } k_t^2 < k^{+2} \end{cases} \right\},
$$

(4.61)

where k^+ represents the crossover point according to Fig. 4.14, and is defined by

$$
k^{+2} = \frac{2m_1 m_2}{\hbar^2 (m_2 - m_1)} \Delta W_c.
$$

(4.62)

Inserting all the points in Z_1 with positive velocity into (4.56) yields the set of states in $\mathbf{k}_1$-space for which a transition is possible. The result is

$$
T_1^+ = \left\{ (k_{1x}, \mathbf{k}_t): \begin{cases} k_{1x} > 0 & \text{if } k_t^2 > k^{+2} \\[2ex] k_{1x} > \sqrt{\dfrac{m_2 - m_1}{m_2}(k^{+2} - k_t^2)} & \text{if } k_t^2 < k^{+2} \end{cases} \right\}.
$$

(4.63)

T_1^+ is indicated by the hatched area in Fig. 4.15. Note that in contrast to the case $m_2 < m_1$ (see Fig. 4.7) a transition is even possible with nearly zero k_{1x}, once the transverse wave vector exceeds k^+.

Instead of using the energy origin of region 1, we can repeat the analysis with the origin in V_1 shifted by ΔW_c, thus obtaining the transition region in $\mathbf{k}_2$-space from (4.43). The figure of Z_2 is the same as for Z_1 in Fig. 4.14, but with zero energy at the minimum of the V_2^*-parabola. By shifting the energies in (4.61) we obtain

$$
Z_2 = \left\{ (W_2, \mathbf{k}_t): \begin{cases} W_2 \geq \dfrac{\hbar^2}{2m_1} k_t^2 - \Delta W_c & \text{if } k_t^2 \geq k^{+2} \\[2ex] W_2 \geq \dfrac{\hbar^2}{2m_2} k_t^2 & \text{if } k_t^2 < k^{+2} \end{cases} \right\}
$$

(4.64)

for Z_2, in accordance with (4.43). Inserting the points in Z_2 with negative velocity into the inverse of W_2

$$
K_2(W_2, \mathbf{k}_t) = \pm \sqrt{\frac{2m_2}{\hbar^2} W_2 - k_t^2}
$$

(4.65)

gives the transition region T_2^- in $\mathbf{k}_2$-space:

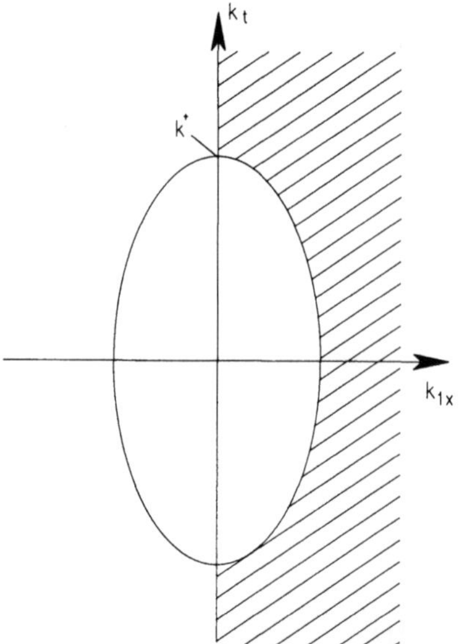

Fig. 4.15 Transition region in $\mathbf{k}_1$-space

$$
T_2^- = \left\{ (k_{2x}, \mathbf{k}_t) \colon \begin{cases} k_{2x} < -\sqrt{\dfrac{m_2 - m_1}{m_1}(k_t^2 - k^{+2})} & \text{if } k_t^2 > k^{+2} \\[2mm] k_{2x} < 0 & \text{if } k_t^2 < k^{+2} \end{cases} \right\}.
$$

$$(4.66)$$

This region is sketched in Fig. 4.16. Only electrons in states of the hatched area are able to transit to region 1. In Fig. 4.16 we notice another remarkable fact: Although the potential energy in region 2 is higher than in region 1, there are possible states of electrons which are not allowed to fall down the step! Once electrons in region 2 have gained enough velocity parallel to the interface, they are not able to fall back to region 1, but are reflected at the interface. Not much attention has been paid in the literature to this effect (however, see e.g. [78]), and to my knowledge, devices making use of this effect have not yet been investigated.

4.6.4 The Interface Condition for General Bands

Now we are in a position to recast the interface conditions in the generalized formalism of the preceding paragraphs. While we have been a bit sloppy in Section 4.2 with the ability of transforming $\mathbf{k}_2$ into $\mathbf{k}_1$ and vice versa, we explicitly want to distinguish here the cases of possible or impossible transition. Starting from (4.16)–(4.17), and regarding the conditions

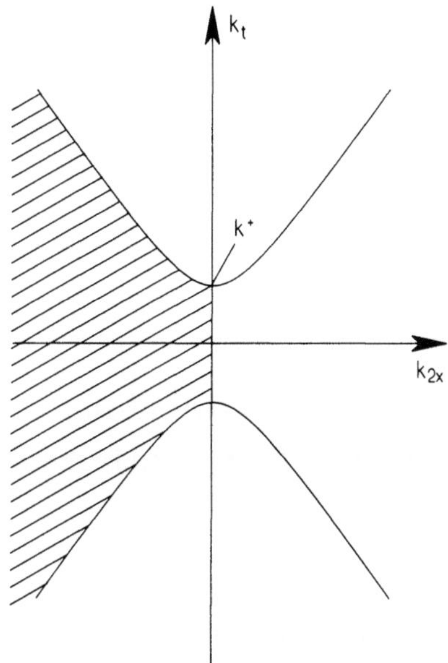

Fig. 4.16 Transition region in $\mathbf{k}_2$-space

Table 4.2. *Interface conditions for the Boltzmann equation with general band structures*

$$
v_{2x}(\mathbf{k}_2)f_2(\mathbf{k}_2)\,d^3\mathbf{k}_2 = v_{1x}(\mathbf{k}_1)D^+(\mathbf{k}_1)f_1(\mathbf{k}_1(\mathbf{k}_2))\,d^3\mathbf{k}_1
$$
$$
\qquad\qquad - v_{2x}(\mathbf{k}_2')[1 - D^-(\mathbf{k}_2')]f_2(\mathbf{k}_2'(\mathbf{k}_2))\,d^3\mathbf{k}_2'
$$
$$
(\mathbf{k}_2 \in T_2^+,\ \mathbf{k}_2' \in T_2^-,\ \mathbf{k}_1 \in T_1^+) \tag{4.67}
$$
$$
v_{2x}(\mathbf{k}_2)f_2(\mathbf{k}_2)\,d^3\mathbf{k}_2 = -v_{2x}(\mathbf{k}_2')f_2(\mathbf{k}_2'(\mathbf{k}_2))\,d^3\mathbf{k}_2'
$$
$$
(\mathbf{k}_2 \in B_2^+,\ \mathbf{k}_2' \in B_2^-) \tag{4.68}
$$
$$
v_{1x}(\mathbf{k}_1)f_1(\mathbf{k}_1)\,d^3\mathbf{k}_1 = v_{2x}(\mathbf{k}_2)D^-(\mathbf{k}_2)f_2(\mathbf{k}_2(\mathbf{k}_1))\,d^3\mathbf{k}_2
$$
$$
\qquad\qquad - v_{1x}(\mathbf{k}_1')[1 - D^+(\mathbf{k}_1')]f_1(\mathbf{k}_1'(\mathbf{k}_1))\,d^3\mathbf{k}_1'
$$
$$
(\mathbf{k}_1 \in T_1^-,\ \mathbf{k}_1' \in T_1^+,\ \mathbf{k}_2 \in T_2^-) \tag{4.69}
$$
$$
v_{1x}(\mathbf{k}_1)f_1(\mathbf{k}_1)\,d^3\mathbf{k}_1 = -v_{1x}(\mathbf{k}_1')f_1(\mathbf{k}_1'(\mathbf{k}_1))\,d^3\mathbf{k}_1'
$$
$$
(\mathbf{k}_1 \in B_1^-,\ \mathbf{k}_1' \in B_1^+) \tag{4.70}
$$

(4.49)–(4.50) for transition, the complete collection of conditions is obtained as given in Table 4.2. Eqs. (4.67)–(4.68) consider the inflow into region 2, where the former contains states with both transition and reflection, while the latter gathers the states for which a transition is not possible and which thus have been reflected in any case. Eqs. (4.69)–(4.70) constitute the inflow into region 1, and are fully symmetric to (4.67)–(4.68). Although not explicitly noted everywhere, it is understood that $\mathbf{k}_1$ and $\mathbf{k}_2'$ depend on $\mathbf{k}_2$ in (4.67)–(4.68), and that $\mathbf{k}_2$ and $\mathbf{k}_1'$ depend on $\mathbf{k}_1$ in (4.69)–(4.70). The relations of $\mathbf{k}_1$, $\mathbf{k}_1'$, $\mathbf{k}_2$, $\mathbf{k}_2'$ are given by (4.51), (4.52), and (4.55), respectively.

As in Subsection 4.2.3, the notation of the interface condition can be simplified by cancelling common factors with the help of (A.1) and (A.12). Note that Appendix A shows that these relations are valid for arbitrary band structures.

Eqs. (4.67)–(4.70) then reduce to

$$f_2(\mathbf{k}_2) = D^+(\mathbf{k}_1(\mathbf{k}_2))f_1(\mathbf{k}_1(\mathbf{k}_2))$$
$$+ \left[1 - D^-(\mathbf{k}_2'(\mathbf{k}_2))\right]f_2(\mathbf{k}_2'(\mathbf{k}_2))$$
$$(\mathbf{k}_2 \in T_2^+, \mathbf{k}_2' \in T_2^-, \mathbf{k}_1 \in T_1^+), \tag{4.71}$$

$$f_2(\mathbf{k}_2) = f_2(\mathbf{k}_2'(\mathbf{k}_2))$$
$$(\mathbf{k}_2 \in B_2^+, \mathbf{k}_2' \in B_2^-), \tag{4.72}$$

$$f_1(\mathbf{k}_1) = D^-(\mathbf{k}_2(\mathbf{k}_1))f_2(\mathbf{k}_2(\mathbf{k}_1))$$
$$+ \left[1 - D^+(\mathbf{k}_1'(\mathbf{k}_1))\right]f_1(\mathbf{k}_1'(\mathbf{k}_1))$$
$$(\mathbf{k}_1 \in T_1^-, \mathbf{k}_1' \in T_1^+, \mathbf{k}_2 \in T_2^-), \tag{4.73}$$

$$f_1(\mathbf{k}_1) = f_1(\mathbf{k}_1'(\mathbf{k}_1))$$
$$(\mathbf{k}_1 \in B_1^-, \mathbf{k}_1' \in B_1^+). \tag{4.74}$$

Again it has to be emphasized that the nature of flux balancing is no longer visible in (4.71)–(4.74), and the original formulation is conceptually to be preferred.

It can be seen from (4.71)–(4.74) that, in fact, the interface condition does not explicitly depend on the energy band structure. That this must be the case becomes clear after considering the special case of thermodynamic equilibrium. In this case, the distribution functions on both sides are the equilibrium distributions with equal temperatures and Fermi levels, depending only on the energy and being independent of the band structure. This is a very general rule which can be satisfied only if (4.67)–(4.70) can be reduced to (4.71)–(4.74).

Outgoing from the interface conditions for the Boltzmann equation of this section, interface conditions for balance equations can be derived along the same lines as in Sections 4.4 and 4.5.

4.7 Non-Conservative Interface Conditions

So far we have assumed throughout in this chapter that energy and transverse momentum is conserved when an electron crosses the interface or is reflected. As discussed in Section 4.2, this assumption requires that scattering effects inside the interface region can be neglected. In most practical cases, the assumption is feasible. As we will see in the following chapters, the commonly used boundary and interface conditions are normally based on this assumption.

However, exceptions to this rule exist. In the transport theory for Schottky contacts of Crowell and Sze [79], for instance, a probability for scattering during interface transport is included in a special factor of the contact current expression (see Section 6.3). In surface roughness scattering [80] at semiconductor-insulator interfaces, a random distribution of the directions of the reflected carriers is assumed; hence momentum conservation does not hold in this process (see Chapter 8).

Thus, for completeness a generalized form of the interface conditions is given in this section which includes the possibility that energy or transverse momentum are not conserved.

We start from Eqs. (4.16)–(4.17) in Section 4.2, or for general band structures from (4.67) and (4.69) in Section 4.6.

There, the conservation of energy and transverse momentum led to relations $\mathbf{k}_2(\mathbf{k}_1)$ and $\mathbf{k}_1(\mathbf{k}_2)$ between the wave vectors on both sides of the interface. Only transitions were allowed where the initial and final wave vectors obeyed these relationships. Now the situation is different. If scattering in the interface region has to be considered, for a given final $\mathbf{k}$ every other wave vector can have been in principle the initial wave vector, depending on the scattering probability. To this end, we have to introduce generalized transition and reflection probabilities $\tilde{D}$ and $\tilde{R}$, respectively. The probabilities $\tilde{D}$ and $\tilde{R}$ now include both the quantum mechanical transition probabilities and the scattering probabilities. The notation is

$$\tilde{D}^+(\mathbf{k}_1', \mathbf{k}_2))\, d^3\mathbf{k}_2 = \text{transition probability from } \mathbf{k}_1' \text{ to } \mathbf{k}_2$$

and

$$\tilde{R}^-(\mathbf{k}_2', \mathbf{k}_2))\, d^3\mathbf{k}_2 = \text{reflection probability from } \mathbf{k}_2' \text{ to } \mathbf{k}_2.$$

The probabilities for the opposite direction are defined analogously. The relationships of the various wave vectors and probabilities are schematically depicted in Fig. 4.17.

The probabilities $\tilde{D}^+$ and $\tilde{R}^+$ resp. $\tilde{D}^-$ and $\tilde{R}^-$ are not independent of each other. The conservation of particle number requires that each impinging electron must go somewhere. This means that the integration of the total probability over the wave vectors after scattering must yield unity:

$$\int_{v_{2x}(\mathbf{k}_2)>0} d^3\mathbf{k}_2\, \tilde{D}^+(\mathbf{k}_1', \mathbf{k}_2) + \int_{v_{1x}(\mathbf{k}_1)<0} d^3\mathbf{k}_1\, \tilde{R}^+(\mathbf{k}_1', \mathbf{k}_1) = 1, \quad (4.75)$$

$$\int_{v_{1x}(\mathbf{k}_1)<0} d^3\mathbf{k}_1\, \tilde{D}^-(\mathbf{k}_2', \mathbf{k}_1) + \int_{v_{2x}(\mathbf{k}_2)>0} d^3\mathbf{k}_2\, \tilde{R}^-(\mathbf{k}_2', \mathbf{k}_2) = 1. \quad (4.76)$$

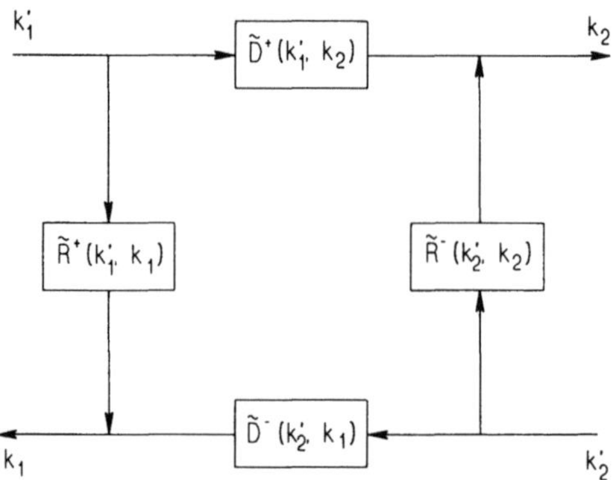

Fig. 4.17 Relations of wave vectors and scattering probabilities

Also the conditions of microscopic reversibility and detailed balance (reciprocity) in equilibrium have to be satisfied by the probabilities [14].
Now we are going to formulate the interface conditions for the Boltzmann equation again as the particle inflow into each of the regions at a certain wave vector. Because of the scattering, we have to integrate over all possible wave vectors that describe an approach towards the interface. We obtain

$$
v_{2x}(\mathbf{k}_2)f_2(\mathbf{k}_2) = \int_{v_{1x}(\mathbf{k}'_1)>0} d^3\mathbf{k}'_1\, v_{1x}(\mathbf{k}'_1)\tilde{D}^+(\mathbf{k}'_1, \mathbf{k}_2)f_1(\mathbf{k}'_1)
$$
$$
- \int_{v_{2x}(\mathbf{k}'_2)<0} d^3\mathbf{k}'_2\, v_{2x}(\mathbf{k}'_2)\tilde{R}^-(\mathbf{k}'_2, \mathbf{k}_2)f_2(\mathbf{k}'_2), \tag{4.77}
$$
$$
v_{1x}(\mathbf{k}_1)f_1(\mathbf{k}_1) = \int_{v_{2x}(\mathbf{k}'_2)<0} d^3\mathbf{k}'_2\, v_{2x}(\mathbf{k}'_2)\tilde{D}^-(\mathbf{k}'_2, \mathbf{k}_1)f_2(\mathbf{k}'_2)
$$
$$
- \int_{v_{1x}(\mathbf{k}'_1)>0} d^3\mathbf{k}'_1\, v_{1x}(\mathbf{k}'_1)\tilde{R}^+(\mathbf{k}'_1, \mathbf{k}_1)f_1(\mathbf{k}'_1), \tag{4.78}
$$

where the primed $\mathbf{k}$'s denote the wave vectors with velocities directed towards the interface, and the unprimed $\mathbf{k}$ are the inflowing (with respect to the regions) wave vectors. Conditions (4.77)–(4.78) should be compared to (4.16)–(4.17).
If Eqs. (4.77)–(4.78) are the interface conditions generalized for the case of non-conservation of energy and transverse momentum, it must be possible to recover the conservative result as a special case. In order to achieve this, we have to set

$$
\tilde{D}^+(\mathbf{k}'_1, \mathbf{k}_2) = \delta(\mathbf{k}_2 - \mathbf{k}_2(\mathbf{k}'_1)) \cdot D^+(\mathbf{k}'_1) \tag{4.79}
$$

and

$$\tilde{R}^-(\mathbf{k}_2', \mathbf{k}_2) = \delta(\mathbf{k}_2 - \mathbf{k}_2(\mathbf{k}_2'))[1 - D^-(\mathbf{k}_2')], \tag{4.80}$$

where D^+ and D^- are the quantum-mechanical transition probabilities introduced in Section 4.2. For the opposite direction, we have

$$\tilde{D}^-(\mathbf{k}_2', \mathbf{k}_1) = \delta(\mathbf{k}_1 - \mathbf{k}_1(\mathbf{k}_2')) \cdot D^-(\mathbf{k}_2') \tag{4.81}$$

and

$$\tilde{R}^+(\mathbf{k}_1', \mathbf{k}_1) = \delta(\mathbf{k}_1 - \mathbf{k}_1(\mathbf{k}_1'))[1 - D^+(\mathbf{k}_1')], \tag{4.82}$$

respectively. The conservation of energy and transverse crystal momentum is achieved as before by the wave vector relations $\mathbf{k}_2(\mathbf{k}_1')$ resp. $\mathbf{k}_1(\mathbf{k}_2')$ for transmission and $\mathbf{k}_2(\mathbf{k}_2')$ resp. $\mathbf{k}_1(\mathbf{k}_1')$ for reflection. The transmission probabilities D^+ resp. D^- are undefined at wave vectors where the transition is forbidden; hence no contribution to the integrals occurs in that case. By insertion of the above probabilities into (4.75) and (4.76), it can be easily verified that the normalization conditions are satisfied.

Substituting (4.79) into (4.77), we get for the first term on the right-hand side

$$\int d^3k_1' \, v_{1x}(\mathbf{k}_1')\delta(\mathbf{k}_2 - \mathbf{k}_2(\mathbf{k}_1'))D^+(\mathbf{k}_1')f_1(\mathbf{k}_1')$$

$$= \int d^3k_2' \, v_{2x}(\mathbf{k}_2')\delta(\mathbf{k}_2 - \mathbf{k}_2')D^+(\mathbf{k}_1(\mathbf{k}_2'))f_1(\mathbf{k}_1(\mathbf{k}_2'))$$

$$= v_{2x}(\mathbf{k}_2)D^+(\mathbf{k}_1(\mathbf{k}_2))f_1(\mathbf{k}_1(\mathbf{k}_2)), \tag{4.83}$$

where, when going from the top line of (4.83) to the next one, we transformed the integration variable from $\mathbf{k}_1'$ to $\mathbf{k}_2'$ by $\mathbf{k}_2' = \mathbf{k}_2(\mathbf{k}_1')$ with the help of (4.18). It is understood that the integration domains comprise only those wave vectors where transitions from region 1 to region 2 are allowed. Likewise, substitution of (4.80) into (4.77) gives for the second term

$$-\int_{v_{2x}(\mathbf{k}_2')<0} d^3k_2' \, v_{2x}(\mathbf{k}_2')\delta(\mathbf{k}_2 - \mathbf{k}_2(\mathbf{k}_2'))[1 - D^-(\mathbf{k}_2')]f_2(\mathbf{k}_2')$$

$$= \int_{v_{2x}(\mathbf{k}_2'')>0} d^3k_2'' \, v_{2x}(\mathbf{k}_2'')\delta(\mathbf{k}_2 - \mathbf{k}_2'')[1 - D^-(\mathbf{k}_2'(\mathbf{k}_2''))]$$

$$\times f_2(\mathbf{k}_2'(\mathbf{k}_2''))$$

$$= v_{2x}(\mathbf{k}_2)[1 - D^-(\mathbf{k}_2'(\mathbf{k}_2))]f_2(\mathbf{k}_2'(\mathbf{k}_2)). \tag{4.84}$$

Here, the integration variable has been transformed by $\mathbf{k}_2'' = \mathbf{k}_2(\mathbf{k}_2')$ to the inflow wave vector. Note that the sign changes during the transformation from outflow to inflow wave vectors because of (A.12) in Appendix A.

With the last line of each of Eqs. (4.83) and (4.84) inserted in (4.77), a common factor $v_{2x}(\mathbf{k}_2)$ can be cancelled out of (4.77), and the resulting equation becomes equal to (4.71). If $D^+ = D^- = 0$, i.e. reflection only, (4.72) results instead. Since (4.71)–(4.72) have been shown to be equivalent to the interface conditions (4.67)–(4.68), we find that indeed the conservative interface condition is recovered by introducing the probabilities (4.79)–(4.80). The condition for inflow into the opposite direction, i.e. (4.69)–(4.70), is recovered analogously by insertion of (4.81) and (4.82) into (4.78).

4.8 Multiple Bands and Interface States

Now we are going to extend the considerations of the previous section to the general case that several bands have to be taken into account in each region. In the simplest case, these are the conduction and valence bands of the semiconductor at the interface, but also more complicated band structures like the upper valleys of GaAs can be described by the formulation. The electron transport in multiple bands can be formulated by specifying a separate distribution function for each of the bands. A set of Boltzmann equations is written down for each distribution function, which are coupled by interband scattering terms (see Eq. (2.7)). For the moment it is advantageous not to distinguish between conduction and valence bands. Hence we consider each band as a concept to define the allowed states for *electrons*, where we recognize valence bands by their negative effective masses.

In order to fix the notation, we put an index onto the distribution function that refers to the corresponding energy band:

$$f_v(\mathbf{k}_v) = \text{distribution function of energy band } v.$$

It is convenient for the following to drop our use of numbers 1 and 2 to denote the regions to the left resp. right of the interface; instead we adopt the meaning that

$$\mu \text{ denotes an energy band on the left-hand side } (x < x_0),$$

and

$$v \text{ denotes an energy band on the right-hand side } (x > x_0).$$

As long as we considered only a single band on each side, the possible transport processes at a semiconductor interface have simply been either transition or reflection. The presence of multiple bands, however, gives rise to a number of additional possibilities, like intervalley transfer at the interface, or surface generation and recombination. Since these processes are in general inelastic, it is convenient to extend the formulation of non-conservative interface transport of Section 4.7 to transitions between several bands. To this end, we supplement the transition and reflection probabilities

$\tilde{D}$ and $\tilde{R}$ with band indices that indicate the source and the target bands for the respective process. The notation is

$$\tilde{D}^{+}_{\mu',\nu}(\mathbf{k}_{\mu'}, \mathbf{k}_{\nu})\, d^3\mathbf{k}_{\nu} = \text{transition probability from band } \mu' \text{ to band } \nu$$

and

$$\tilde{R}^{-}_{\nu',\nu}(\mathbf{k}_{\nu'}, \mathbf{k}_{\nu})\, d^3\mathbf{k}_{\nu} = \text{reflection probability from band } \nu' \text{ to band } \nu.$$

Note that according to our index convention the transition probability $\tilde{D}^{+}_{\mu',\nu}$ describes a transfer in the positive x-direction (from left to right), while $\tilde{R}^{-}_{\nu',\nu}$ describes a process where an electron in band ν' on the right-hand side is reflected, and leaves the interface in band ν.

The normalization conditions of the many-band probabilities are fully analogous to (4.75)–(4.76), with an additional summation over the target energy bands.

The interface conditions are similar to (4.77)–(4.78), supplemented by summations over the energy bands of origin. We find

$$v_{\nu x}(\mathbf{k}_\nu)f_\nu(\mathbf{k}_\nu) = \sum_{\mu'} \int_{v_{\mu'x}(\mathbf{k}_{\mu'})>0} d^3\mathbf{k}_{\mu'}\, v_{\mu'x}(\mathbf{k}_{\mu'})\tilde{D}^{+}_{\mu'\nu}(\mathbf{k}_{\mu'}, \mathbf{k}_\nu)f_{\mu'}(\mathbf{k}_{\mu'})$$
$$- \sum_{\nu'} \int_{v_{\nu'x}(\mathbf{k}_{\nu'})<0} d^3\mathbf{k}_{\nu'}\, v_{\nu'x}(\mathbf{k}_{\nu'})\tilde{R}^{-}_{\nu'\nu}(\mathbf{k}_{\nu'}, \mathbf{k}_\nu)f_{\nu'}(\mathbf{k}_{\nu'}),$$

$$(4.85)$$

$$v_{\mu x}(\mathbf{k}_\mu)f_\mu(\mathbf{k}_\mu) = \sum_{\nu'} \int_{v_{\nu'x}(\mathbf{k}_{\nu'})<0} d^3\mathbf{k}_{\nu'}\, v_{\nu'x}(\mathbf{k}_{\nu'})\tilde{D}^{-}_{\nu'\mu}(\mathbf{k}_{\nu'}, \mathbf{k}_\mu)f_{\nu'}(\mathbf{k}_{\nu'})$$
$$- \sum_{\mu'} \int_{v_{\mu'x}(\mathbf{k}_{\mu'})>0} d^3\mathbf{k}_{\mu'}\, v_{\mu'x}(\mathbf{k}_{\mu'})\tilde{R}^{+}_{\mu'\mu}(\mathbf{k}_{\mu'}, \mathbf{k}_\mu)f_{\mu'}(\mathbf{k}_{\mu'}).$$

$$(4.86)$$

With these equations, in principle every possible transition process between multiple bands in the regions separated by the interface can be described, provided the respective probabilities are known. It should be noted at this place that the probabilities of transitions between the same bands in both regions, e.g. Γ-Γ-transitions, are in general much greater than for transitions between different bands, like e.g. Γ-L-transitions [81, 82].

The interaction of conduction electrons with interface states has been excluded from our considerations so far. However, as has been mentioned in Section 3.2, these states can lead to enhanced generation and recombination effects at the interface. Hence, in the following we will furthermore include transitions to and from interface states (traps) in the above analysis. We consider a number of discrete interface states, labeled with the index t. Each of the traps lies at an energy level W_t. If we prefer a continous description by the interface state density $D_t(W_t)$, as introduced in Chapter 3, the sum over the interface states below has to be replaced by an integral $\int D_t(W_t)\,dW_t$.

The traps at each energy have a surface density N_t, measured in cm^{-2}. A typical value for N_t is 10^{12} cm^{-2} [28]. The occupation of the traps by electrons are described by the probability f_t. It follows that the surface density of the *occupied* states is $N_t f_t$.

The interface states interact with the conduction and valence bands; electrons can be captured by an empty state and emitted by an occupied state. In addition to the interaction of the traps with the energy bands, we have to consider transitions between the traps themselves. These processes are described by respective transition probabilities

$$D_{v't}(\mathbf{k}_{v'}) = \text{probability that an electron in band } v' \text{ is captured,}$$

$$D_{tv}(\mathbf{k}_v)\,d^3\mathbf{k}_v = \text{probability rate that an electron is released to band } v,$$

Table 4.3. *Interface conditions for multiple bands and interface states*

$$
\begin{aligned}
v_{vx}(\mathbf{k}_v)f_v(\mathbf{k}_v) = {} & \sum_{\mu'} \int_{v_{\mu'x}(\mathbf{k}_{\mu'})>0} d^3\mathbf{k}_{\mu'}\, v_{\mu'x}(\mathbf{k}_{\mu'})\tilde{D}^+_{\mu',v}(\mathbf{k}_{\mu'}, \mathbf{k}_v)f_{\mu'}(\mathbf{k}_{\mu'}) \\
& - \sum_{v'} \int_{v_{v'x}(\mathbf{k}_{v'})<0} d^3\mathbf{k}_{v'}\, v_{v'x}(\mathbf{k}_{v'})\tilde{R}^-_{v',v}(\mathbf{k}_{v'}, \mathbf{k}_v)f_{v'}(\mathbf{k}_{v'}) \\
& + \sum_t D_{tv}(\mathbf{k}_v)N_t f_t \tag{4.87}
\end{aligned}
$$

$$
\begin{aligned}
v_{\mu x}(\mathbf{k}_\mu)f_\mu(\mathbf{k}_\mu) = {} & \sum_{v'} \int_{v_{v'x}(\mathbf{k}_{v'})<0} d^3\mathbf{k}_{v'}\, v_{v'x}(\mathbf{k}_{v'})\tilde{D}^-_{v'\mu}(\mathbf{k}_{v'}, \mathbf{k}_\mu)f_{v'}(\mathbf{k}_{v'}) \\
& - \sum_{\mu'} \int_{v_{\mu'x}(\mathbf{k}_{\mu'})>0} d^3\mathbf{k}_{\mu'}\, v_{\mu'x}(\mathbf{k}_{\mu'})\tilde{R}^+_{\mu'\mu}(\mathbf{k}_{\mu'}, \mathbf{k}_\mu)f_{\mu'}(\mathbf{k}_{\mu'}) \\
& - \sum_t D_{t\mu}(\mathbf{k}_\mu)N_t f_t \tag{4.88}
\end{aligned}
$$

$$
\begin{aligned}
N_t \frac{df_t}{dt} = {} & \sum_{\mu'} \int_{v_{\mu'x}(\mathbf{k}_{\mu'})>0} d^3\mathbf{k}_{\mu'}\, v_{\mu'x}(\mathbf{k}_{\mu'})D_{\mu't}(\mathbf{k}_{\mu'})f_{\mu'}(\mathbf{k}_{\mu'}) \\
& - \sum_{v'} \int_{v_{v'x}(\mathbf{k}_{v'})<0} d^3\mathbf{k}_{v'}\, v_{v'x}(\mathbf{k}_{v'})D_{v't}(\mathbf{k}_{v'})f_{v'}(\mathbf{k}_{v'}) \\
& - N_t f_t \sum_v \int_{v_{vx}(\mathbf{k}_v)>0} d^3\mathbf{k}_v\, D_{tv}(\mathbf{k}_v) \\
& - N_t f_t \sum_\mu \int_{v_{\mu x}(\mathbf{k}_\mu)<0} d^3\mathbf{k}_\mu D_{t\mu}(\mathbf{k}_\mu) + \sum_{t'} [D_{t't}N_{t'}f_{t'} - D_{tt'}N_t f_t] \tag{4.89}
\end{aligned}
$$

and

$$D_{tt'} = \text{probability rate that an electron jumps from trap } t \text{ to trap } t.$$

It has to be noted that $D_{tv}(\mathbf{k}_v)$ is the probability per time interval and k-space volume element that the electron is emitted from the trap; its units are $\mathrm{s}^{-1}\mathrm{m}^3$. Similarly, $D_{tt'}$ is the probability per time interval of the transition between traps; it is measured in s^{-1}.

The inflows into bands v and μ are now set up according to (4.85) and (4.86), respectively, complemented by the stream of carriers that are emitted from the traps into the respective band. The result is listed in Table 4.3 in Eq. (4.87)–(4.88). The number of the carriers released from traps is computed by the product of the transition probability and the density of the occupied interface states.

The occupation probability f_t is determined by the balance of electrons being captured and being released in (4.89). The first two terms in this equation are the contribution of the electrons captured from all the considered bands; the third and fourth term account for the emitted electrons. The last two terms represent the electrons interchanged with other trap levels t'.

The normalization conditions for the occurring probabilities are found in analogy to (4.75)–(4.76). The summation over the states after scattering now additionally includes the sum over the target bands and the sum over the trap states. We obtain

$$\sum_v \int_{v_{vx}(\mathbf{k}_v)>0} d^3\mathbf{k}_v \, \tilde{D}^+_{\mu'v}(\mathbf{k}_{\mu'}, \mathbf{k}_v)$$

$$+ \sum_\mu \int_{v_{\mu x}(\mathbf{k}_\mu)<0} d^3\mathbf{k}_\mu \, \tilde{R}^+_{\mu'\mu}(\mathbf{k}_{\mu'}, \mathbf{k}_\mu) + \sum_t D_{\mu't}(\mathbf{k}_{\mu'})$$

$$= 1, \tag{4.90}$$

$$\sum_\mu \int_{v_{\mu x}(\mathbf{k}_\mu)<0} d^3\mathbf{k}_\mu \, \tilde{D}^-_{v'\mu}(\mathbf{k}_{v'}, \mathbf{k}_\mu)$$

$$+ \sum_v \int_{v_{vx}(\mathbf{k}_v)>0} d^3\mathbf{k}_v \, \tilde{R}^-_{v'v}(\mathbf{k}_{v'}, \mathbf{k}_v) + \sum_t D_{v't}(\mathbf{k}_{v'})$$

$$= 1. \tag{4.91}$$

Still, the probabilities have to satisfy the reciprocity conditions [14].

Eqs. (4.87)–(4.89) represent the most general interface conditions for the Boltzmann equation. Conditions for the balance equations can be derived from these along the lines pointed out in Section 4.4 by the inflow moments method. The above formulation is able to describe electron transport at semiconductor interfaces including the effects of thermionic emission [83], elastic tunneling [83], phonon-assisted (inelastic) tunneling [84], quantum-mechanical reflection [85], inelastic reflection (backscattering by phonons [79]), interband transfer [81], surface recombination and generation [86], tunneling-assisted recombination [87], trap-assisted tunneling, surface trapping [88] and interface state interaction. All these effects are hidden in the various probabilities, which of course must be known in order to use the equations in a practical application. Several special cases from this collection will be presented in the subsequent chapters. A qualitative discussion of the most important effects can be found in [28].

Semiconductor-Insulator Interface 5

Semiconductor-insulator interfaces are normally occuring in every semiconductor device because they usually constitute the physical boundary of the device. The devices have a passivating coating (e.g. silicon dioxide for silicon devices), are encapsulated in a plastic case, or are adjacent to surrounding air. In integrated circuits they are sometimes isolated from each other by trenches filled with insulator material. From this it is evident that boundary conditions for semiconductor-insulator interfaces are an indispensable component of any device simulator.

The insulating property of an insulator material is due to its very large band gap which is responsible for very low concentrations of free electrons and hence for a low conductivity. Semiconductor-insulator interfaces are thus characterized by a very large conduction band edge difference ΔW_c (cf. Fig. 5.1). Examples are

$$\text{Si-SiO}_2: \qquad \Delta W_c = 3.2 \text{ eV}$$

$$\text{Si-vacuum:} \quad \Delta W_c = \chi = 4.05 \text{ eV}$$

for interfaces between silicon and silicon dioxide or vacuum, respectively. (Note that for an interface to vacuum, ΔW_c is identical to the electron affinity χ.) Compared to the band gap of silicon of 1.12 eV, and keeping in mind the exponential dependence of the electron concentration on energy differences, these are very large values.

According to the treatment in Section 2.5 we have to distinguish interface conditions from boundary conditions. These alternatives are treated in the following two sections.

5.1 Interface Conditions

Looking at the semiconductor-insulator interface as an interface condition for device simulation means that we want to simulate the electronic behaviour both in the semiconductor and the insulator. Continuity and Poisson equation have to be treated separately in this context.

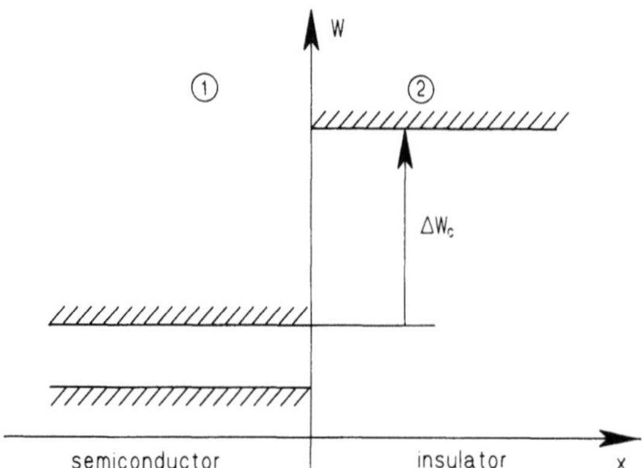

Fig. 5.1 Energy bands at a semiconductor-insulator interface

If we are interested in the interface condition for the *continuity equation*, this means that we maintain in our simulator a distribution of electrons and holes in the semiconductor as well as in the insulator. The interface condition then describes the transition of carriers between semiconductor and insulator. Since the insulator normally has a very low concentration of carriers, this corresponds to *charge emission into the insulator*. To put it the other way round, an interface condition for the continuity equations is necessary only if the emission of electrons or holes from the semiconductor into the insulator has to be taken into account.

Normally, the effect of charge emission into an insulator is not of much interest with respect to semiconductor devices. This means that in usual applications of device simulation the charge concentration in the insulator can be neglected. It then suffices to replace the continuity equations in the insulator by $n = p = 0$. Since this equation is an algebraic one, interface conditions are not necessary, and boundary conditions for the continuity equations in the semiconductor have to be used instead. This case will be addressed in Section 5.2.1.

An exception to this treatment may be necessary for the simulation of MOSFETs. Because the insulator between the gate contact and the semiconductor channel is thin, the knowledge of the carrier concentrations in the gate insulator may become important. Especially for the investigation of MOSFET degradation by injection of hot electrons into the gate oxide, the simulation of gate leakage currents, or the tunneling of electrons through the insulator to the floating gate of EEPROMs, carriers in the insulator have to be considered. These cases require special interface conditions and are not considered here (see for example [41, 39, 89]).

With respect to the *Poisson equation*, the situation is different. Although the electron concentration in the insulator might be negligible, the distribution of the electric potential can be nevertheless important, and the Poisson equation has to be solved in the insulator as well. This leads to the situation that with respect to the continuity equation the semiconductor-insulator interface is treated as a boundary, and with respect to the Poisson equation as an interface. An example is again the MOSFET, where in many cases the gate isolation may be treated as ideally insulating, while the electric field between gate and channel (as the basic functional principle of the MOSFET) has to be computed.

The interface conditions for the *Poisson equation* result from electrostatic considerations and have been introduced in Chapter 3. They are, rewritten for material 2 as the insulator and material 1 as the semiconductor,

$$\varphi_i - \varphi_s = \frac{1}{q}(\chi_s - \chi_i - \Delta W_c), \tag{5.1}$$

$$-\varepsilon_i \frac{\partial \varphi_i}{\partial x} + \varepsilon_s \frac{\partial \varphi_s}{\partial x} = \sigma, \tag{5.2}$$

where ε_i and ε_s are the dielectric constants of the insulator and the semiconductor, respectively, and σ is a possible surface charge at the interface.

5.2 Boundary Conditions

In this section, we consider the case where the simulation domain ends at the interface, and the insulator is regarded as being part of the external world. This means that we need a boundary condition for the differential equations in the semiconductor which expresses the effect of the environment on the semiconductor. In the present chapter, we concentrate on the non-degenerate drift-diffusion transport model as introduced in Chapter 2. The boundary conditions for the continuity equations and the Poisson equations are treated separately in the following subsections.

5.2.1 Continuity Equations

According to Chapter 4, the boundary condition for a continuity equation expresses the inflow of the respective particle kind into the simulation domain. Hence, in this subsection we have to model the flow of electrons at the semiconductor-insulator interface. Taking into account the usual case of one conduction band and one valence band, we have two continuity equations for electrons and holes, respectively. Because we have to account for recombination and generation processes through the interface states in the bandgap, we start the analysis with the interface conditions of Section 4.8.

Referring to the situation depicted in Fig. 5.1, the inflow into the semi-conductor (region 1, bands labeled by μ) is given by (4.88). In the following, we adapt this fairly general formulation to the present case. Eq. (4.87) has to be discarded here, because the inflow into the external region 2 (bands labeled by v) is not considered in a *boundary* condition. As a result, we will obtain a model of recombination and generation at the semiconductor surface. We are going to use the derivation of the Shockley-Read-Hall recombination in the bulk [19], [44] as a guideline; see also the discussion of surface recombination in [86].

Since we consider one conduction and one valence band, the band indices μ or μ' introduced in Section 4.8 can assume only the values n and p. In order to follow the usual analysis of bulk Shockley-Read-Hall recombina-tion [19], we consider only a single trap level at energy W_t.

According to the discussion in Section 2.5, the influence of the environment across a boundary is accounted for by assuming a suitable model of the external distribution function. Because of the extremely low electron con-centration in the insulator, it is acceptable to assume a zero distribution function in the insulator:

$$f_v = 0. \tag{5.3}$$

Hence, the first term on the right-hand side of (4.88) vanishes.

Now we turn to the reflection probability $\tilde{R}^+$. First, we neglect direct inter-actions between the conduction band and the valence band; recombination or generation processes at the interface shall occur only via the interface state. This is expressed by vanishing "non-diagonal" elements of $\tilde{R}^+$:

$$\tilde{R}^+_{np} = \tilde{R}^+_{pn} = 0. \tag{5.4}$$

Second, we have to consider the electron conservation condition (4.90). Because of the very large band-edge discontinuity at the semiconductor-insulator interface, we assume that no electrons can escape the semiconduc-tor by setting

$$\tilde{D}^+_{\mu'v} = 0 \tag{5.5}$$

in (4.90), and also

$$D_{tv} = 0 \tag{5.6}$$

in (4.89). Insertion of (5.4) and (5.5) into (4.90) yields

$$\int_{v_{\mu x}(\mathbf{k}_\mu)<0} d^3 k_\mu \, \tilde{R}^+_{\mu\mu}(\mathbf{k}'_\mu, \mathbf{k}_\mu) + D_{\mu t}(\mathbf{k}'_\mu) = 1. \tag{5.7}$$

The above discussion is schematically summarized in Fig. 5.2. An electron, arriving at the interface in the valence or the conduction band, can either be reflected within the same band or get trapped at the interface state. An

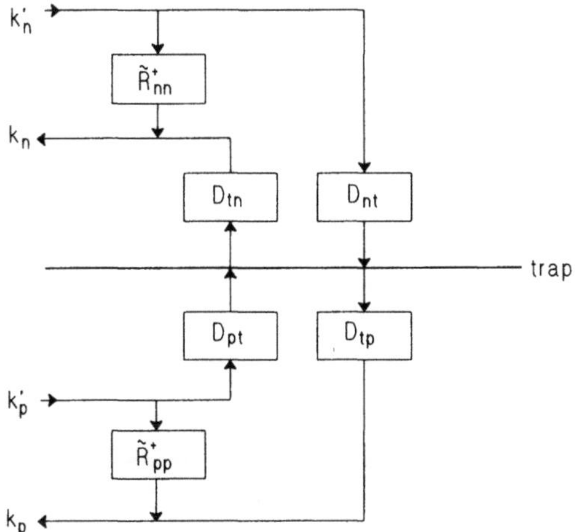

Fig. 5.2 Transition probabilities at the semiconductor-insulator interface

electron in the trap can be released to either the conduction or the valence band. Transmission of electrons into or from the insulator does not occur. The vertical branches on the right-hand side of the figure visualize the trap-assisted recombination and generation processes.

The capture and release probabilities must in any case obey the Pauli principle and the rule of detailed balance. From these considerations, we find

$$D_{\mu t}(\mathbf{k}) = C_\mu(\mathbf{k})N_t[1 - f_t] \tag{5.8}$$

for the capture probability and

$$D_{t\mu}(\mathbf{k}) = -v_{\mu x}(\mathbf{k})C'_\mu(\mathbf{k})[1 - f_\mu(\mathbf{k})] \tag{5.9}$$

for the emission probability. The factor $[1 - f]$ in both expressions accounts for the Pauli principle by making the probability zero if the target state is occupied. Note that the velocity component in (5.9) is negative for states into which electrons can be released from the trap; thus the right-hand side of the equation is always positive. The proportionality factors C_μ and C'_μ are related to each other by the principle of detailed balance:

$$v_{\mu x}(\mathbf{k})C'_\mu(\mathbf{k}) = -v_{\mu x}(-\mathbf{k})C_\mu(-\mathbf{k})\exp\left(\frac{W_t - W_\mu(\mathbf{k})}{k_B T}\right). \tag{5.10}$$

C_n and C'_p, measured in units of cm^2, are the electron and hole capture cross sections [90] (up to a constant factor of about unity). Typical values for capture cross sections are 10^{-15}–10^{-14} cm^2 [28]. The quantity $W_\mu(\mathbf{k})$ in (5.10) is the band structure function of band μ according to the definition in Section 3.1. It is $W_c + \hbar^2 k^2/(2m_n)$ for the conduction band and $W_v - \hbar^2 k^2/(2|m_p|)$ for the valence band.

Inserting (5.3), (5.4) and (5.5) into (4.88), we obtain as the boundary condition (still for the Boltzmann equation) in the present case

$$v_{\mu x}(\mathbf{k}_\mu)f_\mu(\mathbf{k}_\mu) = -\int_{v_{\mu x}(\mathbf{k}'_\mu)>0} d^3\mathbf{k}'_\mu\; v_{\mu x}(\mathbf{k}'_\mu)\tilde{R}^+_{\mu\mu}(\mathbf{k}'_\mu,\,\mathbf{k}_\mu)f_\mu(\mathbf{k}'_\mu)$$
$$- D_{t\mu}(\mathbf{k}_\mu)N_t f_t. \tag{5.11}$$

Physically, the main assumptions leading to (5.11) are equivalent to neglecting both electron concentration in the insulator as well as emission of electrons into the insulator. The former assumption is justified by the small carrier concentration in the insulator, while the latter results from negligible thermionic emission across the large potential step. For semiconductor-insulator interfaces normally considered in device simulations, these assumptions are usually valid. As mentioned in Section 5.1, an important exception is the investigation of gate injection in MOSFETs.

If the Monte Carlo method [7] is used for solving the Boltzmann equation, the above boundary condition can be realized by randomly selecting the wave vector of the returning carrier in accordance with the reflection or emission probabilities. Depending on the roughness of the interface, the reflection probability may be of *specular* or *diffusive* type [14]. Mixed models, i.e. a superposition of specular and diffusive reflection, are also in use [14]. The roughness of the semiconductor-insulator interface is important in the analysis of electron flow in the channel of a MOSFET, because it presents an additional obstruction mechanism to the electrons. This particular case will be considered in more detail in Chapter 8.

The occupation probability f_t of the trap is in principle determined by (4.89). For the present analysis, we assume a quasi-stationary behaviour of the trap occupancy [19] [44]. Setting the left-hand side of (4.89) to zero, we get under the above assumptions

$$\int_{v_{nx}(\mathbf{k}_n)>0} d^3\mathbf{k}_n\; v_{nx}(\mathbf{k}_n)D_{nt}(\mathbf{k}_n)f_n(\mathbf{k}_n)$$
$$+\int_{v_{px}(\mathbf{k}_p)>0} d^3\mathbf{k}_p\; v_{px}(\mathbf{k}_p)D_{pt}(\mathbf{k}_p)f_p(\mathbf{k}_p)$$
$$- N_t f_t\int_{v_{nx}(\mathbf{k}_n)<0} d^3\mathbf{k}_n\; D_{tn}(\mathbf{k}_n)$$
$$- N_t f_t\int_{v_{px}(\mathbf{k}_p)<0} d^3\mathbf{k}_p\; D_{tp}(\mathbf{k}_p)$$
$$= 0, \tag{5.12}$$

where we have made explicit the sums over μ and μ'. Insertion of (5.8)–(5.9) into (5.12) gives

$$N_t[1 - f_t][I_n^+ + I_p^+] - N_t f_t[I_n^- + I_p^-] = 0, \tag{5.13}$$

where the abbreviations

$$I_\mu^+ = \int_{v_{\mu x}(\mathbf{k})>0} d^3k \; v_{\mu x}(\mathbf{k})C_\mu(\mathbf{k})f_\mu(\mathbf{k}) \tag{5.14}$$

and

$$I_\mu^- = -\int_{v_{\mu x}(\mathbf{k})<0} d^3k \; v_{\mu x}(\mathbf{k})C_\mu'(\mathbf{k})[1 - f_\mu(\mathbf{k})] \tag{5.15}$$

have been introduced. Solving (5.13) for f_t gives the occupation probability

$$f_t = \frac{I_n^+ + I_p^+}{I_n^+ + I_p^+ + I_n^- + I_p^-}. \tag{5.16}$$

According to the principle introduced in Section 4.4, the *boundary conditions for the continuity equations* are obtained by integrating (5.11) over the inflow region of $\mathbf{k}_\mu$ for each involved band. The inflow moment for the conduction band thus reads

$$\int_{v_{nx}(\mathbf{k})<0} d^3k \; v_{nx}(\mathbf{k}_n)f_n(\mathbf{k})$$

$$= -\int_{v_{nx}(\mathbf{k})>0} d^3k \; v_{nx}(\mathbf{k}_n)f_n(\mathbf{k})[1 - D_{nt}(\mathbf{k})]$$

$$- N_t f_t \int_{v_{nx}(\mathbf{k})<0} d^3k \; D_{tn}(\mathbf{k}), \tag{5.17}$$

where $\tilde{R}_{nn}^+$ has already been substituted with the help of (5.7). The first term in the brackets can be put on the left-hand side of the equation, where it complements the integration range to the full k-space. With (2.10), the left-hand side then becomes proportional to the electron current density. Hence we can rewrite (5.17) to give

$$J_{nx} = -\frac{qN_t}{4\pi^3}[(1 - f_t)I_n^+ - f_t I_n^-], \tag{5.18}$$

where we again used the abbreviations (5.14)–(5.15). Analogously, we find from the inflow moment of the valence band the corresponding current density

$$J_{px} = -\frac{qN_t}{4\pi^3}[(1 - f_t)I_p^+ - f_t I_p^-]. \tag{5.19}$$

Inserting (5.16), we get

$$J_{px} = -\frac{qN_t}{4\pi^3} \frac{I_n^- I_p^+ - I_n^+ I_p^-}{I_n^+ + I_p^+ + I_n^- + I_p^-} \tag{5.20}$$

for the hole current density, and

$$J_{nx} = \frac{qN_t}{4\pi^3} \frac{I_n^- I_p^+ - I_n^+ I_p^-}{I_n^+ + I_p^+ + I_n^- + I_p^-} \tag{5.21}$$

for the electron current density.

In order to compute the abbreviations defined in (5.14)–(5.15), we have to insert the ansatz for the distribution function. In consistence with the non-degenerate drift-diffusion transport model, we basically have to use f from (2.13) for the electron distribution. For the valence band, we correspondingly use $g(\mathbf{k}) = 1 - f(-\mathbf{k})$ according to (2.8). These assumptions give us J_{nx} and J_{px} as functions of the carrier concentrations and current densities at the surface. In a first order approximation, however, we can neglect the corrections coming from the current densities. With these premises, we obtain

$$\begin{aligned}
I_n^+ &= 4\pi^3 v_{n0} n C_n, \\
I_n^- &= 4\pi^3 v_{n0} n_1 C_n, \\
I_p^+ &= 4\pi^3 v_{p0} p_1 C_p', \\
I_p^- &= 4\pi^3 v_{p0} p C_p',
\end{aligned} \tag{5.22}$$

where we introduced the velocities

$$v_{n0} = \sqrt{\frac{k_B T}{2\pi m_n}}, \tag{5.23}$$

$$v_{p0} = \sqrt{\frac{k_B T}{2\pi |m_p|}} \tag{5.24}$$

and the concentrations

$$n_1 = N_c \exp\left(\frac{W_t - W_c}{k_B T}\right), \tag{5.25}$$

$$p_1 = N_v \exp\left(-\frac{W_t - W_v}{k_B T}\right). \tag{5.26}$$

During the calculation, we used the fact that n and p in non-degeneration are much smaller than the respective effective state densities N_c and N_v. Further, we replaced the $\mathbf{k}$-dependent scattering cross sections C_n and C_p' with constant average values.

Inserting (5.22) into (5.20) and (5.21), we can write the boundary condition in the final form

$$J_{px} = qs_p \frac{np - n_i^2}{n + n_1 + \dfrac{s_p}{s_n}(p + p_1)}, \tag{5.27}$$

$$J_{nx} + J_{px} = 0. \tag{5.28}$$

This is the familiar form of Shockley-Read-Hall (SRH) recombination at the surface [86]. The quantities n and p are the electron and hole concentrations at the interface, respectively. The parameters s_n and s_p of the recombination model have the dimension m s^{-1} and are called electron resp. hole "recombination velocities"; they are given by

$$s_n = N_t C_n v_{n0}, \tag{5.29}$$

$$s_p = N_t C'_p v_{p0}, \tag{5.30}$$

The recombination velocities are connected to the capture cross sections through the occurence of C_n and C'_p in (5.29) and (5.30), respectively [86]. Typical values for the surface recombination velocity are $s_n \approx s_p \approx 10$ cm s^{-1} for high quality surfaces and $s_n \approx s_p \approx 10^4$–$10^5$ cm s^{-1} for surfaces with large defect density [91]. Further parameters of the model are the concentrations n_1 and p_1, which are identically defined as in the volume SRH effect [19]. In (5.27) we used the relation $n_1 p_1 = n_i^2$, following from (5.25)–(5.26), where the intrinsic concentration n_i is defined by [19]

$$n_i = \sqrt{N_c N_v} \exp\left(-\frac{W_c - W_v}{2k_B T}\right). \tag{5.31}$$

It is a material parameter of the semiconductor, independent of the trap energy.

As noted in the beginning, the derivation of the SRH model involves only a single trap level. In reality, we have a distribution of traps across the total bandgap. This could be accounted for by a respective *a posteriori* integration of (5.27) over the trap distribution [90]; however, even then we would disregard the transitions between traps. In addition, the result would become too complicated to be used in device simulations. The common practice is thus to regard (5.27) as an engineering model of surface recombination, where the parameters have to be adapted to experiments.

The second boundary condition, i.e. the vanishing of the total charge current density in (5.28), re-expresses of course the initial assumptions that no carriers cross the interface. We notice from (5.28) that the flow J_{px} of holes towards the interface fully recombines with the flow J_{nx} of electrons.

Several simplifications of (5.27) are in use [42], with (5.28) remaining valid. If n is large, (5.27) can be simplified to the *minority carrier condition* [92]

$$J_{px} = qs_p(p - p_0) \tag{5.32}$$

by taking the asymptotic limit $n \to \infty$. The quantity p_0 is defined by

$$p_0 = \frac{n_i^2}{n}. \tag{5.33}$$

The second simplification is obtained in the case of negligible recombination at the interface. Taking the limit $s_p \to 0$ in (5.27) yields the *zero recombination condition*

$$J_{px} = 0. \tag{5.34}$$

Now the electron and hole continuity equations are decoupled at the interface. In fact, by insertion of (5.34) into (5.28), we see that J_{nx} also vanishes. The reason is, of course, that electrons possibly flowing towards the interface could neither recombine with holes nor transit into the insulator. The combination (5.34) and (5.28) is most commonly used in device simulations at the semiconductor-insulator interface, where surface recombination effects are not expected to be important.

Finally, the case of *strong recombination* can be considered. This is the limit of $s_n, s_p \to \infty$, and from (5.27) follows in this case

$$np = n_i^2. \tag{5.35}$$

Inserting (2.14) and the corresponding relation for holes, we obtain equivalently a condition for the quasi-Fermi levels:

$$\phi_n = \phi_p. \tag{5.36}$$

We note that a surface with strong recombination is sometimes considered as equivalent to an ohmic contact [93]. In Chapter 6 it is shown that mechanisms exist at a contact which are different from the recombination mechanism considered here, but nevertheless lead to a condition which resembles (5.36).

In conclusion of this subsection, we simplify the analysis to the case where a single band is sufficient for the description of electron transport. To be specific, let us take the conduction band. The results can be applied to the case of a single valence band in a straightforward manner.

If only the conduction band needs to be considered, interaction between the trap and the valence band can be neglected; hence $D_{pt} = D_{tp} = 0$ is assumed. Eq. (5.16) then can be simplified to

$$f_t = \frac{I_n^+}{I_n^+ + I_n^-}. \tag{5.37}$$

Inserting this expression into (5.18), we obtain

$$J_{nx} = 0. \tag{5.38}$$

This is the resulting one-band boundary condition for the electron continuity equation at the semiconductor-insulator interface. The condition states that no net electron flow occurs at the interface. It is intuitively plausible, since both transition to the insulator and recombination with holes have been excluded. With (5.22) inserted into (5.37), the trap occupation probability turns out to be nearly unity in this case.

If charge transport in the semiconductor can be described solely using the valence band, the analogous boundary condition is of course

$$J_{px} = 0. \tag{5.39}$$

Since this analysis is restricted to a unipolar model of electrons or holes alone, we have only a single continuity equation; hence one of the equations (5.38) or (5.39) is sufficient as a boundary condition for the respective equation.

5.2.2 Poisson Equation

A boundary condition for the Poisson equation is needed if the electrostatic potential shall be computed only in the semiconductor, but not in the insulator. This is the case at the outer boundary of the semiconductor device. All the interesting electronic processes are occuring in the interior of the semiconductor region, and the field distribution in the surrounding insulator often is of minor interest. For this reason, the boundary of the simulation domain is often placed at the interface between the semiconductor region and the protective insulator.

As a boundary condition for the Poisson equation, a relation for the potential in the semiconductor at the interface is required. Rewriting (5.2) for the potential φ_s gives

$$\frac{\partial \varphi_s}{\partial x} = \frac{1}{\varepsilon_s}\left[\sigma + \varepsilon_i \frac{\partial \varphi_i}{\partial x}\right]. \tag{5.40}$$

However, in order to serve as a boundary condition, the electric field $-\partial \varphi_i/\partial x$ on the outside has to be known. But, since φ_i is not part of the simulation, it is unknown and must be replaced by an assumption. It turns out that the conditions normally used in the device simulators are simply

$$\frac{\partial \varphi_s}{\partial x} = \frac{\sigma}{\varepsilon_s} \tag{5.41}$$

or even

$$\frac{\partial \varphi_s}{\partial x} = 0. \tag{5.42}$$

In (5.41), the electric field in the insulator is assumed to be zero; the field inside the semiconductor is balanced by a surface charge σ. In (5.42), a zero surface charge is assumed in addition.

The justification for simplifying (5.40) to (5.42) in the literature of device simulation is usually not very pronounced. In many cases, (5.42) is stated without further explanation. Reiser [94, 95]—in one of the earliest papers on device simulation—only notes that the assumption (5.42) brings forth a great simplification of matters. He further observes that (5.42) results if $\varepsilon_i = 0$ (and $\sigma = 0$). However, what we can expect at best is that ε_s is much larger than ε_i. But under this condition (5.40) only reduces to (5.42) if $\sigma = 0$ and the normal component of the outside electric field is not exceeding the field in the semiconductor, the latter argument again waiting for justification.

Let us consider each term on the right-hand-side of (5.40) separately. If the surface charge is a known constant, or a model for it is available, the σ-term represents no problem. Otherwise, $\sigma = 0$ may be assumed. The significance of this choice can be confirmed by making some test calculations with other reasonable values of σ. Typical values for σ/q at Si-SiO$_2$-interfaces are in the range of 10^{10}–10^{12} cm^{-2}.

The field term may be discussed in the following way. Normally, semiconductor devices as a whole do not carry a net charge. Thus, when looking at the device from far away, it appears uncharged, and consequently does not emit an electric field. If we transfer the condition of vanishing electric field to the semiconductor surface as in (5.42), we neglect the effect of internal charge displacements on the field distribution in the vicinity of the device.

Hence, the general recommendation may be the following. If appreciable electric fields in the vicinity of the device are to be expected, one should place the boundary of the simulation region somewhat more outside where setting the electric field to zero is reasonable. After that, the correct interface condition (5.2) can be used at the semiconductor-insulator interface, and the potential and field distribution in the surrounding insulator is computed from the Poisson equation. On the boundary of the insulator, in turn, the boundary condition (5.42) may be used.

5.3 Virtual Boundaries

Due to the always limited computer resources, in device simulation the simulation domain has to be selected in such a way that the domain is as small as possible, but as large as necessary in order to include all the effects that are important for the solution of the actual problem. Often this results in the specification of boundaries which have no physical origin at all, but rather constitute an artificial clipping of the real situation.

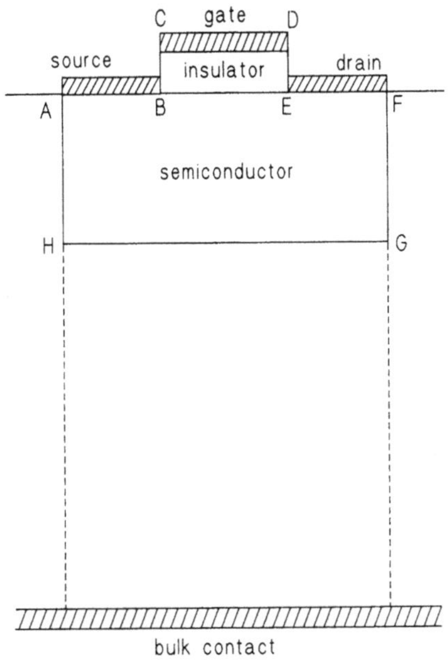

Fig. 5.3 Simulation domain for a MOSFET device

A typical structure for device simulation is depicted in Fig. 5.3 (after [48]),
wich was the basic simulation geometry of the early versions of the widely
known MINIMOS simulator [96]. The figure shows the geometry of a
MOSFET together with the boundaries used for the simulation. While the
boundary sections A-B, E-F (ohmic contact) and B-E (semiconductor-insulator interface) represent physical boundaries, the sections F-G, G-H and
H-A have been chosen arbitrarily in order to limit the size of the simulation
domain. The position of the *virtual boundaries* must be selected such that
the simulated electronic processes inside the device are not perturbed by
the choice. This means that the virtual boundaries must be placed in positions where "nothing happens", i.e. where both currents and normal electric field are negligible, and no influence from the environment is present.
Thus, the boundary condition for the Poisson equation on a virtual boundary is again the vanishing normal electric field condition of (5.42). With
respect to the electron resp. hole continuity equations, the boundary
conditions (5.38) and (5.39), respectively, are used. In case of the inclusion
of energy balance equations, the respective boundary conditions are vanishing normal components of the energy current densities at the boundary.
These boundary conditions have the effect of a total decoupling of the
simulation region from the environment on the other side of the boundary.
It is clear that the introduction of this boundary condition is only valid if
the boundary is far away from the active regions of the device. In cases

where this is not possible—for instance if the neighbouring device is too close—a virtual boundary between the two devices is not allowed because a significant interaction between the two devices is to be expected. Under circumstances like these, either the coupled devices have to be simulated as a whole, or the interaction of the device with the surrounding region must be modelled in an approximate way, e.g. by a lumped network approximation.

Metal-Semiconductor Contact   6

Contact models have been playing a role in device simulation from its early beginnings [97], [98], [94], [95], because every device needs contacts for the application of voltages and the provision of electric currents. Both low resistance (ohmic) and rectifying (Schottky) contacts are thus essential parts of many semiconductor devices. Accordingly, models for ohmic as well as Schottky contacts have been established in many device simulation programs. These models are introduced as boundary conditions for the numerical solution of the semiconductor transport equations.

In the first two sections of the present chapter, the general features of a metal-semiconductor contact are discussed, and a comprehensive contact model is developed. Boundary conditions for the limiting cases of Schottky and ohmic contacts will be derived by corresponding simplifications in the sections to follow.

6.1 Band Model and Current Transport Mechanisms

6.1.1 Band Model

The band model of a metal-semiconductor contact is shown schematically in Fig. 6.1 for the case of an *n*-conducting semiconductor. Normally, in equilibrium a depletion region develops in the semiconductor at the interface [30]. Hence, in case of electron conduction the bands bend upwards as shown in Fig. 6.1, forming a barrier for the flow of electrons. The *barrier height* is defined as

$$\phi_B = W_{cS} - \phi_M, \tag{6.1}$$

where ϕ_M is the Fermi energy in the metal and W_{cS} is the conduction band edge at the peak of the barrier (see Fig. 6.1). The barrier height is nearly independent of doping density. Note that a constant barrier height implies that the carrier concentrations at the interface in equilibrium are also inde-

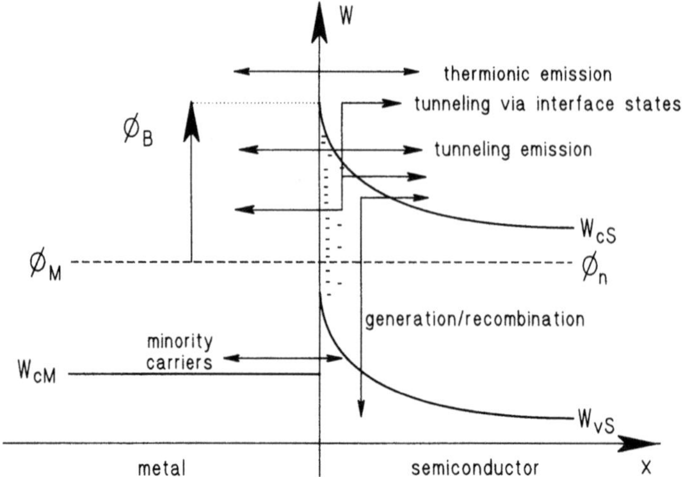

Fig. 6.1 Energy bands at a metal-semiconductor contact

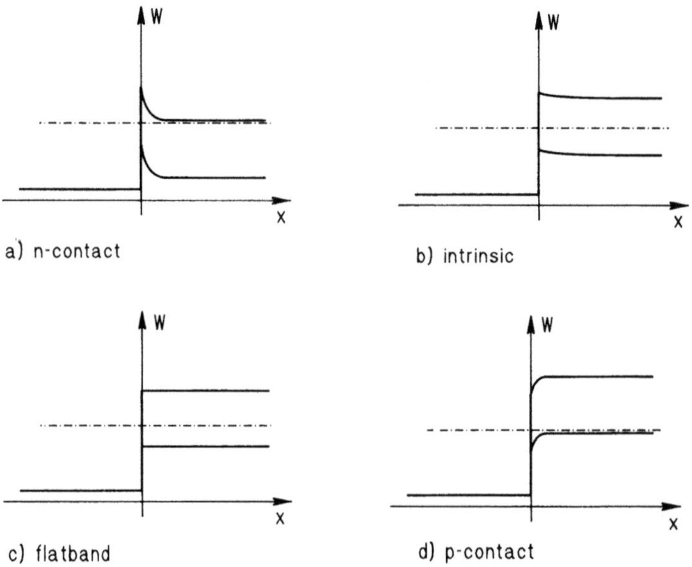

Fig. 6.2 Metal contact for donor and acceptor doping

pendent of doping density, because the barrier height fixes the position of
the Fermi level in the bandgap.
Fig. 6.2 shows what happens if we decrease the donor doping until zero
(intrinsic case), and increase henceforth the acceptor doping until we arrive
at a strongly *p*-conducting semiconductor. The dot-dashed line is the equilibrium Fermi level.

Fig. 6.2a depicts the case of an n-contact, as in Fig. 6.1. In contrast to the electrons, the holes see a trough (or "pocket") at the interface where they tend to accumulate (recall that the holes have a tendency to move upwards in the band diagram).

The situation in the intermediate region from low donor doping over the intrinsic case to low acceptor doping depends on the relation of barrier height ϕ_B and bandgap energy W_g. If $\phi_B > Wg/2$, as shown in Fig. 6.2, the accumulation of holes in the "pocket" at the interface dominate the space charge as soon as the donor doping drops below the (fixed) hole density. At zero doping (Fig. 6.2b) this charge alone is responsible for the rise of the bands in order to satisfy the barrier height condition. (For a discussion of metal contacts on intrinsic and inverted semiconductors, see also [99].)

Increasing now the acceptor doping reduces the band bending further until the flatband case in Fig. 6.2c is reached.

Additional increase of p-doping moves the bulk energy bands further upwards, and a depletion region with a negative space charge develops. As shown in Fig. 6.2d, the bands now bend downwards, and form a barrier with respect to the flow of holes, which have become the majority carriers by this time.

If $\phi_B < Wg/2$ on the contrary, the flatband case is reached before the intrinsic case. Now the electrons accumulate in a pocket and provide for the (now downward) band bending between the flatband and intrinsic cases.

Summarily, we find that in any case for reasonably doped semiconductors a barrier for the majority carriers is present at the interface, which obstructs their free flow into the direction of the metal.

Theoretically, cases are possible where the majority carriers accumulate in the semiconductor at the interface to the metal, leading to band bending in the opposite direction as in Fig. 6.2. However, the case of majority carrier accumulation seems to be very uncommon in practice [30], except for extremely low doping as in Fig. 6.2b. Thus, in this book we will concentrate merely on the depletion case.

In Chapter. 3, we characterized the shift of the band structures on both sides of the interface by the difference of the lowest conduction bands ΔW_c, i.e.

$$\Delta W_c = W_{cS} - W_{cM} \tag{6.2}$$

in this case. With respect to metal-semiconductor contacts, however, it is common practice to use the barrier height ϕ_B (6.1) for the characterization of the interface instead. With (6.2), the conversion of ΔW_c to ϕ_B is given by

$$\phi_B = \Delta W_c - (\phi_M - W_{cM}). \tag{6.3}$$

Note that the term in parentheses on the right-hand side of (6.3) is a material specific constant of the metal, because it is related to the metallic electron concentration. This is why the interface can be characterized likewise by ϕ_B instead of ΔW_c.

In like manner, it is convenient to replace the electron affinity of the metal by its work function Φ, i.e. the difference between the macro potential $-q\varphi$ and the Fermi level. It is then possible to write down the equivalent of the electron affinity rule for the metal-semiconductor contact, which gives the barrier height as $\phi_B = \Phi - \chi_S$. However, also this relation is in general not valid! The reason is of course the same as for the failure of the electron affinity rule (see Chapter 3 and [47]).

Unfortunately, the assumption of very large susceptibility α, which led to Eq. (3.22), does not seem to hold for metal-semiconductor contacts [31]; hence a formula as simple as (3.22) cannot be given for the barrier height. Thus it is preferrable for metal-semiconductor contacts to rely on experimental values.

In zero order approximation, ϕ_B is a constant, dependent on the combination of the materials at the contact. Comprehensive tables of barrier heights for a number of materials can be found in [42] and [28]. An example collection is presented in Table 6.1.

We note in passing that a metallic contact on an insulator follows in principle the same rules as a metal-semiconductor contact. The definition of a respective barrier height is fully analogous. The barrier height of metal-insulator interfaces is an important parameter in the band model of MIS-structures; it plays a role for the threshold voltage of MOSFETs. For this reason, also the barrier heights of some metals on silicon dioxide are given in the table, taken from [42].

The experimental values of barrier heights usually show some fluctuation, because they are sensitive to the crystallic orientation and details of the technological processing. The barrier heights in the table thus represent only typical values; for accurate quantitative simulations the barrier height should be determined specifically for the actual process under consideration.

A more detailed analysis shows that ϕ_B depends on the electric field strength in the semiconductor at the interface [30]. The reasons for this dependence can be a thin, insulating interface layer (causing a voltage drop at the layer),

Table 6.1. *Experimental electron barrier heights (in eV)*

	Si	GaAs	AlAs	SiO$_2$
Ag	0.78	0.88		4.2
Al	0.72	0.8		3.2
Au	0.80	0.9	1.2	4.1
Cu	0.58	0.82		3.8
Ni	0.61			3.7
Mo	0.68			
Pt	0.9	0.84	1.0	
PtSi	0.84			

metal-induced gap states (cf. Chapter 3), or the image force lowering effect. The most important of these effects is the last one [99]. The image force leads to a lowering of the barrier height ϕ_B if the electric field is high [30, 42]. The barrier lowering is given by [30]

$$\Delta\phi_B = q \sqrt{\frac{q|E_{max}|}{4\pi\varepsilon_s}}, \tag{6.4}$$

where E_{max} is the maximum electric field in the barrier. The barrier peak is shifted at the same time from $x = 0$ by a very small distance (about 1 nm) into the semiconductor [92]. The effective barrier height then is determined by

$$\phi_B = \phi_{B_0} - \Delta\phi_B, \tag{6.5}$$

where ϕ_{B_0}, the "flatband barrier height" [100], is a parameter depending on the material combination. The barrier lowering effect becomes significant in high reverse bias or high doping. It leads to an increase of the current through the contact.

With regard to device simulation, our point of view is of macroscopic nature to that extent that we assume ΔW_c resp. ϕ_B as a given parameter (either as a constant or as a more complicated model). On this scale microscopic features of the contact, like an insulating interface layer, or the smoothness of transition from metal to semiconductor, are no longer visible. According to the concept developed in Chapters 3 and 4, we characterize the contact by models for the barrier height ϕ_B and the transition probability, and investigate the resulting transport behaviour.

6.1.2 Transport Mechanisms

The prominent mechanisms of carrier transfer at the metal-semiconductor are also depicted in Fig. 6.1. These are

- thermionic emission over the top of the barrier,
- tunneling emission by direct tunneling through the barrier,
- tunneling via interface states (trap-assisted tunneling),
- interface generation and recombination,
- minority carrier transport.

The most important current transport mechanisms for metal-semiconductor contacts in semiconductor devices are *thermionic emission* and *tunneling*. Whether one or the other effect is dominant near equilibrium depends primarily on the doping density.

Figure 6.3 shows the band model in the extreme cases of low doping (a) and high doping (b). The curvature of the barrier depends on the space charge in the depletion region and hence on the doping of the semiconductor:

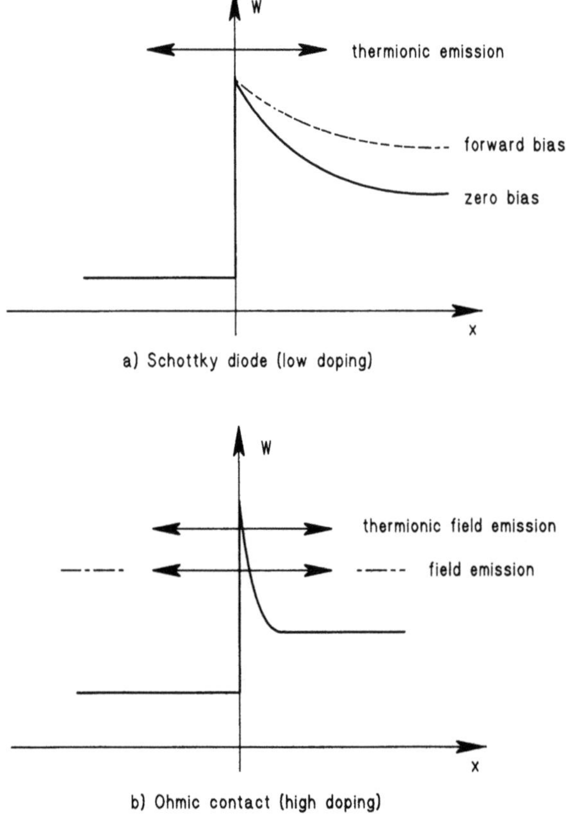

Fig. 6.3 Energy bands at Schottky and ohmic contacts

the higher the doping, the higher is the bending of the bands according to Poisson's equation.

Electron transport in the low doping case is dominated by thermionic emission over the top of the barrier, as indicated. In the case of forward current flow, the apparent barrier height as seen from the semiconductor side is lowered by a raise of the band edge due to the applied bias. This is shown schematically by the dashed line in Fig. 6.3. Hence in forward bias, thermionic emission favours the transition of carriers from the semiconductor to the metal against the opposite direction. Thus, if thermionic emission is the dominant transport mechanism, the current through the contact is strongly dependent on the sign of the contact voltage drop. The contact then is said to have a rectifying behaviour. We will call a contact like this *Schottky contact*[1]. Thermionic emission requires carriers with a high kinetic energy

---

[1] However, note that a number of authors in literature call every interface between metal and semiconductor a Schottky contact, regardless of its rectifying or ohmic behaviour.

in order to be able to surmount the barrier. Because at higher temperatures the number of carriers with high energies increases, thermionic emission is more pronounced if the carrier temperature is rised (that is where the name comes from), or if the barrier height is lowered.

It is obvious from Fig. 6.3 that the thickness of the barrier must decrease if we increase the doping, keeping the barrier height constant. Thus, the high doping extreme is shown in Fig. 6.3b.

From quantum mechanics follows that the probability of carrier tunneling through a potential barrier increases rapidly if the barrier thickness becomes sufficiently small. Thus, above a certain doping level, the tunneling current through the barrier exceeds the thermionic emission current across the barrier, and the characteristics of the contact changes drastically. If the maximum portion of the tunnel current occurs at an energy around the Fermi level, the process is called *field emission*. If the maximum number of tunneling electrons is between the Fermi level and the top of the barrier, the process is called *thermionic field emission*.

The names come from the emission of electrons from a metal or a semiconductor into vacuum if a very high electric field is applied. This effect is depicted in Fig. 6.4. A theoretical description of field-emission from semiconductors has been given in [101]. The high field in the vacuum results in a strong decrease of the potential energy in the vicinity of the metal surface. This in turn makes the barrier which keeps the electrons inside the metal very thin, and at a certain threshold of the field strength electrons start to tunnel out of the metal, i.e. are emitted into the vacuum. Since in a metal the current carrying electrons have an energy around the Fermi level, the maximum number of emitted electrons are in this energy range. It is because of this fact that the term "field emission" is used in the semiconductor case, too. If the temperature of the metal in Fig. 6.4 is rised, more states of

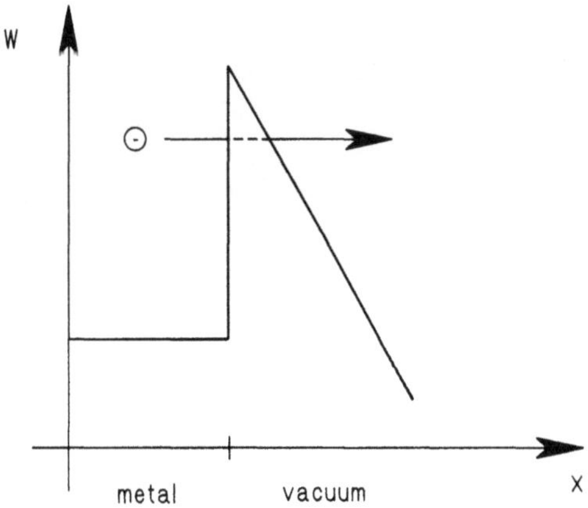

Fig. 6.4 Field emission of electrons from a metal into vacuum

higher energy become populated in the metal, and the maximum no longer occurs at the Fermi level, but is shifted towards the barrier peak. For this reason the situation is called *thermally enhanced field emission* or briefly *thermionic field emission*. In metal-semiconductor contacts, the distinction between field emission and thermionic field emission has little significance, and we will summarily refer to both processes as *tunnel emission*.

The tunneling current can be many orders of magnitude higher than the thermionic emission current. Very small voltage drops across the contact can give rise to very high currents. In other words, the contact has a very small resistivity. Also, the nonlinear rectifying effect is—although theoretically present—no longer noticeable in practice; the current-voltage characteristics appears linear in the limits below which the contact is not damaged by high currents. A contact like this is called *ohmic contact*. Ohmic metal-semiconductor contacts are fabricated by deposition of metal on a highly doped (more than 10^{19} cm^{-3}) semiconductor. They are used to connect the device to the electric circuit.

6.2 Non-Ideal Contact

This section presents a model for the currents through metal-semiconductor contacts including both thermionic emission and tunneling effects. The model allows the simulation of devices with non-ideal contacts where commonly used boundary conditions are not applicable. It also serves as a starting point for the derivation of models for ohmic and Schottky contacts in subsequent sections as specialized cases of the general model. It is a fairly new model which has been published only recently [102, 103].

The starting point of the model is the energy band diagram of the contact in Fig. 6.5. The metal-semiconductor interface is located at $x = 0$, with the metal at $x < 0$ and the semiconductor at $x \geq 0$. The barrier height is ϕ_B, while W_0 is the top energy of the barrier. The Fermi level is indicated by the dot-dashed line. At the interface, the Fermi level is discontinuous; it thus accounts for a voltage drop V_c across the contact, defined by

$$qV_c = \phi_n - \phi_M, \tag{6.6}$$

where ϕ_M is the Fermi level of the metal, and ϕ_n the quasi-Fermi level of the semiconductor.

As indicated in the figure, tunneling through the barrier occurs within an energy range W_T below the top of the barrier. Hence tunneling takes place within a distance x_T from the metal junction, where x_T is the maximum tunnel length. Since tunneling is not included in the drift-diffusion equations describing electron transport in the semiconductor volume, we place the actual boundary for the simulation domain at the edge x_T of the tunneling region. Thus, the device simulator may keep with semi-classical trans-

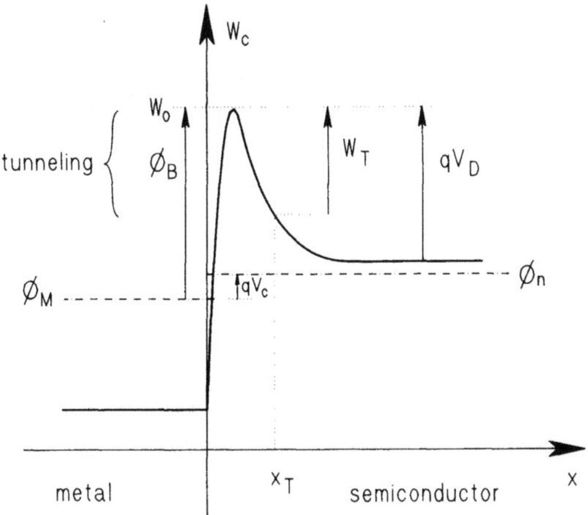

Fig. 6.5 Energy band diagram of metal-semiconductor contact

port and does not have to deal with quantum mechanics, while the boundary condition accounts for the tunneling process. W_T and the corresponding x_T will be determined by the condition that tunneling can be neglected for $x > x_T$. However, as apparent from Fig. 6.5, W_T cannot grow further once it reaches qV_D, which corresponds to the situation that the whole barrier is transparent to electrons and can be tunneled through.

While in the course of Eqs. (6.2), (6.1), and (6.3) the interface region was understood to extend up to the conduction band edge maximum, the top of the barrier is now located inside the interface region, (cf. also Fig. 4.1). Accordingly, W_{cS} occuring in the subsequent derivation is the conduction band edge at $x = x_T$:

$$W_{cS} = W_c(x_T). \tag{6.7}$$

The tunneling energy range W_T then is

$$W_T = W_0 - W_c(x_T), \tag{6.8}$$

and the barrier height becomes according to Fig. 6.5

$$\phi_B = W_0 - \phi_M, \tag{6.9}$$

replacing (6.1).

In the following subsections, the contact model will be developed. The goal is to obtain consistent boundary conditions for the drift-diffusion transport model at metal-semiconductor contacts. Boundary conditions for the continuity equations are derived by introduction of approximations for the quantum-mechanical transmission probability and the electron distribution functions. These are required to evaluate the general conditions in Section 4.4 for the present case. In the beginning, we consider only unipolar transport with electrons as majority carriers. Minority carriers and the bipolar model will be discussed in Subsection 6.2.7. The boundary condition for the Poisson equation will be derived in Subsection 6.2.5 from a consideration of the band edge W_c at the boundary.

6.2.1 Transmission Probability

The contact model is based on a one-dimensional analysis of the tunneling transmission probability, which has been given by Crowell and Rideout in [83]. The semiconductor conduction band edge according to [83] is assumed to be parabolic,

$$W_c(x) = W_0 + qE_{\max}x + \frac{q^2 N_D^*}{2\varepsilon_s}x^2, \tag{6.10}$$

where $E_{\max}$ is the maximum electric field at the contact, and qN_D^* is the effective space charge responsible for the band bending in the x-direction. In one dimension and in depletion, N_D^* is the donor doping concentration. Assumption (6.10) is discussed in [83, 104, 99]. Image force effects are accounted for by a respective lowering of ϕ_B [30, 42], see Eq. (6.5). The voltage V_D appearing in Fig. 6.5 is formally defined by the maximum electric field at the interface

$$qV_D = \frac{\varepsilon_s E_{\max}^2}{2N_D^*}; \tag{6.11}$$

it corresponds to the potential difference between the barrier peak and the point where $E_x = 0$. The assumption (6.10) for the shape of the barrier in the computation of the tunneling probability allows tunneling—in the scope of the model—to occur only in the space charge region.

A qualitative picture of the quantum-mechanical transmission probability as a function of electron energy (computed by numerical techniques e.g. in [105]) is drawn in Fig. 6.6. The remarkable feature is that even for energies higher than the barrier peak W_0 the transmission probability is smaller than unity, while classically one would expect that every electron with an energy higher than the peak will cross the interface. Quantum-mechanically however, a fraction of these electrons is reflected. We discussed this quantum-mechanical reflection effect already in Section 4.2.1.

A large number of sophisticated theoretical investigations of tunneling at metal-semiconductor contacts exist, e.g. [30, 104, 83, 106, 107]. However, these either do not fully cover all necessary modes of operation or require cost intensive numerical integrations, which makes them far too expensive for device simulation purposes. For device simulation a simple, explicit formula is needed which comprises the essential physical effects but is easy to evaluate. Moreover, in view of the many assumptions involved anyway,

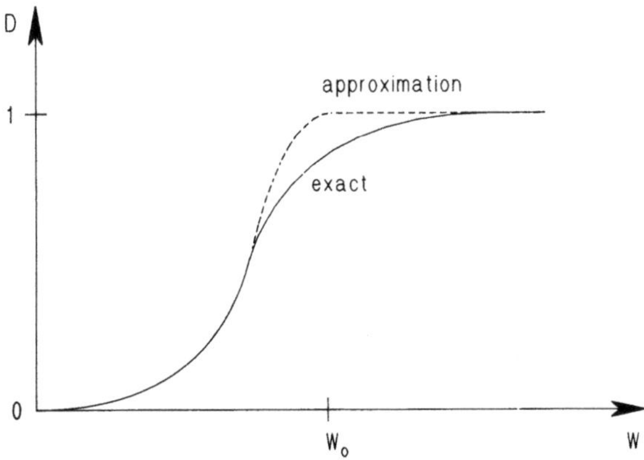

Fig. 6.6 Transmission probabilities as a function of electron energy

a sophisticated treatment of tunneling does not seem to be justified [92, 99].

In our model, we follow Crowell and Rideout in using the *Wentzel-Kramers-Brillouin approximation* (WKB approximation) [108]. The WKB approximation neglects quantum-mechanical reflection, and deviates from the exact solution mainly in the energy range around the barrier peak, as shown in Fig. 6.6.

In the WKB approximation, the tunneling probability for an electron with energy $W < W_0$ is given by

$$D(W) = \exp\left(-2\int_0^{x_0} \sqrt{\frac{2m^*}{\hbar^2}(W_c(x) - W)}\,dx\right), \tag{6.12}$$

where m^* is the tunneling effective mass. It is assumed that m^* is not a function of energy [83]. Values of m^* in silicon for different orientations are discussed in [109]. The location x_0 in (6.12) is the point where the electron of energy W "comes out of the tunnel" (cf. Fig. 6.7), i.e. $W_c(x_0) = W$.

The transmission probability has been computed by Crowell and Rideout using (6.12) and (6.10) to give [83]

$$D(W) = \exp\left(-\frac{\varepsilon_s E_{max}^2}{2N_D^* W_{00}}[\sqrt{c} - (1 - c)\,\text{Artanh}(\sqrt{c})]\right), \tag{6.13}$$

where

$$W_{00} = \frac{q\hbar}{2}\sqrt{\frac{N_D^*}{m^*\varepsilon_s}} \tag{6.14}$$

and

$$c = \frac{2N_D^*}{\varepsilon_s E_{max}^2}(W_0 - W) \tag{6.15}$$

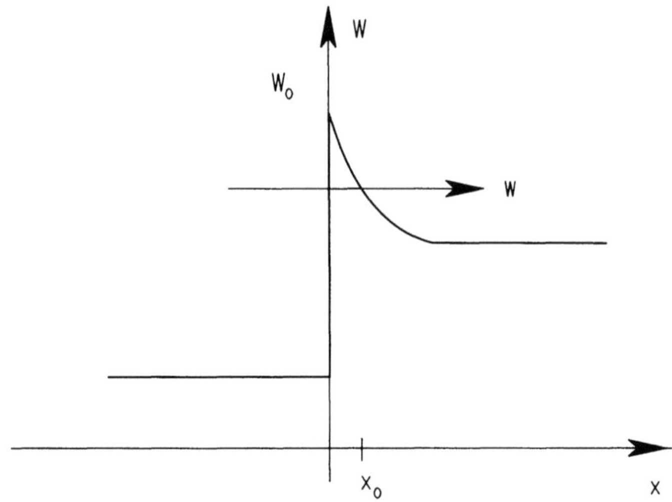

Fig. 6.7 Determination of the tunneling probability

have been introduced as abbreviations. For energies $W > W_0$, a transmission probability of unity is assumed, leading to classical thermionic emission [30]. Crowell and Rideout investigated the contact current using (6.13) in a numerical integration. For device simulation purposes, however, the transmission probability has to be simplified in order to arrive at an analytically integrable expression.

Fig. 6.8 shows the expression in square brackets of (6.13) as a function of c. The c-dependence shows a nearly parabolic shape. For comparison, the graph of c^2 is also displayed in Fig. 6.8. It can be seen from Fig. 6.8 that

$$\sqrt{c} - (1 - c)\,\text{Artanh}(\sqrt{c}) \approx c^2 \tag{6.16}$$

represents a reasonable approximation of the dependence. Despite the fact that the approximation is made in the exponent in (6.13), the approach is chosen since it has the advantage that it leads to a closed formula that is suited for device simulation purposes.

Inserting (6.16) into (6.13), the transmission probability becomes

$$D(W) \approx \exp\left(-\frac{\varepsilon_s E_{max}^2}{2N_D^* W_{00}}\,c^2\right). \tag{6.17}$$

Making use of the abbreviations (6.15) and (6.11), we arrive at

$$D(W) \approx \exp\left(-\frac{(W_0 - W)^2}{W_{00}\,q V_D}\right). \tag{6.18}$$

This approximation will be used subsequently for an analytical evaluation of the tunneling current. It should be noted that the transmission probabil-

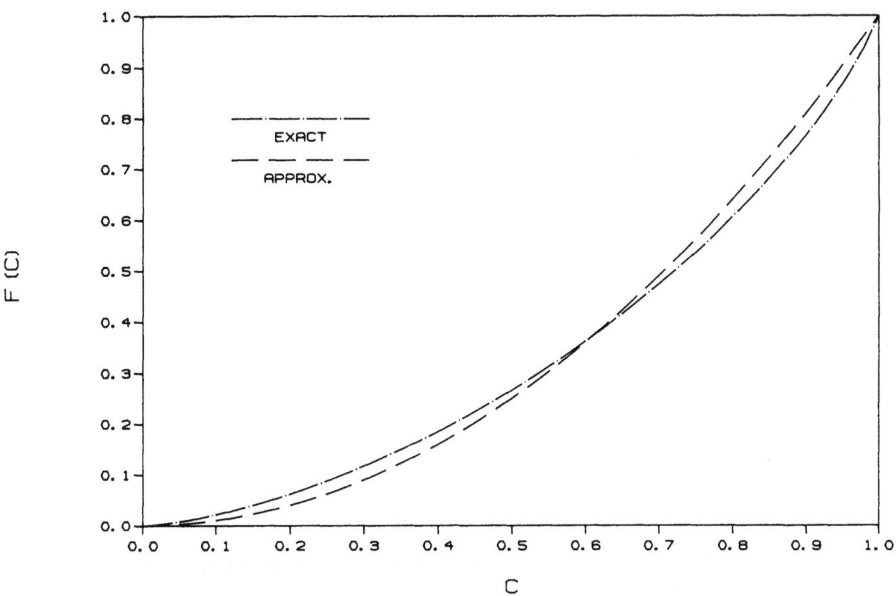

Fig. 6.8 Exact and approximate expression in the tunneling formula

ity is valid for both directions of electron traversal across the interface with energy W.

The above analysis has been a one-dimensional one. For device simulation, however, a three-dimensional description is necessary. An additional difficulty comes into play by allowing distinct effective masses on the two sides of the interface, i.e. in the metal and the semiconductor, whereas a single constant mass has been assumed in the above analysis. If we reconsider the problem in 3D space but with a 1D symmetry, i.e. invariant with respect to translations parallel to the interface, we find that the transmission probability becomes a function of W_{2x} instead of W (see Appendix B),

$$D(W) \rightarrow D(W_{2x}),\tag{6.19}$$

where W_{2x} is defined by

$$W_{2x} = \frac{\hbar^2 k_{2x}^2}{2m_n} + W_{cS}.\tag{6.20}$$

In (6.20), m_n is the effective mass and W_{cS} is the band edge of the semiconductor conduction band at the boundary, respectively. The transition (6.19) from 1D to 3D is valid if the tangential part of the wave function is continous at the interface [101], i.e. the y- and z-dependences are plane waves $\exp(jk_y y + jk_z z)$, going continuously through the interface. As shown in Appendix B, this is the case under the condition that the potential and the effective masses are constant in directions parallel to the interface. A discontinuity of both the potential and the effective mass at the interface is allowed. Additional quantum reflections resulting from the mass discontinuity (see Appendix B) are neglected as well. For simulation purposes, we somewhat relax the condition concerning the potential by demanding that the potential has a sufficiently slow variation parallel to the interface.

Since the barrier is entirely on the semiconductor side of the junction, it is the tunneling effective mass of the *semiconductor* that has to be used in the tunneling probability formula [78].

In summary, the complete expression for the transmission probability for the metal-semiconductor becomes

$$
\boxed{
\begin{aligned}
D^+&(\mathbf{k}_1(\mathbf{k}_2)) \\
&= D^-(-k_{2x}, k_{2y}, k_{2z}) \overset{!}{=} D(k_{2x}) \\
&= \begin{cases}
\exp\left(-\dfrac{(W_{2x} - W_0)^2}{qV_D W_{00}}\right), & \text{if } W_{2x} < W_0 \text{ (tunneling)}, \\[2mm]
1, & \text{if } W_{2x} \geq W_0 \text{ (therm. emission)}.
\end{cases}
\end{aligned}
}
$$

$$(6.21)$$

The formula includes both transition directions. It also includes energies above and below the peak of the barrier, corresponding to thermionic emission and tunneling, respectively. Note that the transmission probability (6.21) does not depend on k_{2y} and k_{2z}, which enables us to write simply $D(k_{2x})$ in the following.

6.2.2 Setting Up the Boundary Condition

In this subsection, the model for the non-ideal metal-semiconductor contact will be set up as a boundary condition for the drift-diffusion transport model. As explained in detail in Chapter 4, a boundary condition for the semiconductor region means taking the inflow into region 2, the semiconductor region in Fig. 6.5. According to (4.16), the boundary condition for the Boltzmann equation in the semiconductor is

$$
\begin{aligned}
v_{2x} f_2(\mathbf{k}_2)\, d^3 k_2 = {}& v_{1x} D^+(\mathbf{k}_1(\mathbf{k}_2)) f_1(\mathbf{k}_1(\mathbf{k}_2))\, d^3 k_1 \\
& + v_{2x}[1 - D^-(-k_{2x}, k_{2y}, k_{2z})] \\
& \times f_2(-k_{2x}, k_{2y}, k_{2z})\, d^3 k_2 \\
& (v_{2x} > 0).
\end{aligned}
$$

$$(6.22)$$

The drift-diffusion transport model involves the moments of order zero and one of the Boltzmann equation (Chapter 2). From Section 4.4 then follows that the appropriate boundary condition is derived with $M(\mathbf{k}) = 1$, i.e. from the moment

$$\int_0^\infty dk_{2x} \int_{-\infty}^\infty dk_{2y}\, dk_{2z} \quad (\langle\text{boundary condition}\rangle), \tag{6.23}$$

where Eq. (6.22) has to be inserted into the integral. The result is

$$\int_0^\infty dk_{2x} \int_{-\infty}^\infty dk_{2y}\, dk_{2z}\, v_{2x} f_2(\mathbf{k}_2)$$

$$- \int_0^\infty dk_{2x} \int_{-\infty}^\infty dk_{2y}\, dk_{2z}\, v_{2x} f_2(-k_{2x}, k_{2y}, k_{2z})$$

$$= \int_0^\infty dk_{2x} \int_{-\infty}^\infty dk_{2y}\, dk_{2z}\, v_{2x} D(k_{2x})$$

$$\times [f_1(\mathbf{k}_1(\mathbf{k}_2)) - f_2(-k_{2x}, k_{2y}, k_{2z})], \tag{6.24}$$

where we collected all terms containing the transition probability on the right-hand side. Also, we indicated in (6.24) that the transition probability depends only on the x-component of the wave vector; see (6.21).

Eq. (6.24) can be further simplified. If we change the sign of the integration variable k_{2x} of the second integral, (6.24) becomes

$$\int_0^\infty dk_{2x} \int_{-\infty}^\infty dk_{2y}\, dk_{2z}\, v_{2x} f_2(\mathbf{k}_2)$$

$$+ \int_{-\infty}^0 dk_{2x} \int_{-\infty}^\infty dk_{2y}\, dk_{2z}\, v_{2x} f_2(\mathbf{k}_2)$$

$$= \int_0^\infty dk_{2x} \int_{-\infty}^\infty dk_{2y}\, dk_{2z}\, v_{2x} D(k_{2x})$$

$$\times [f_1(\mathbf{k}_1(\mathbf{k}_2)) - f_2(-k_{2x}, k_{2y}, k_{2z})]. \tag{6.25}$$

The integrals on the left-hand side combine to give, after multiplication with a factor $-q/(4\pi^3)$, the electron current density according to (2.10). We thus obtain as the metal-semiconductor contact boundary condition for the electron continuity equation the result

$$\boxed{\begin{aligned} J_{nx} &= -\frac{q}{4\pi^3} \int_0^\infty dk_{2x}\, v_{2x} D(k_{2x}) \int_{-\infty}^\infty dk_{2y}\, dk_{2z} \\ &\quad \times [f_1(\mathbf{k}_1(\mathbf{k}_2)) - f_2(-k_{2x}, k_{2y}, k_{2z})]. \end{aligned}} \tag{6.26}$$

An equivalent formulation of this equation can be obtained by changing the k_{2x}-integration variable to the kinetic energy with the help of (4.7) [110]. Also it is possible to change the integration variable from $\mathbf{k}_2$ to $\mathbf{k}_1$

using (4.18). However, care must be taken in this case; the integration domain in region 1 has a more complicated shape (see Fig. 4.7), since a large number of metal electrons is reflected at the interface. Further, the velocity in front of D changes from v_{2x} to v_{1x} because of (4.18), which is not regarded in the literature in any case [111].

According to the procedure outlined in Section 4.4, we now have to choose appropriate functional forms for the distribution functions f_1 and f_2 in (6.26).

The *metal* is considered as a reservoir of electrons with a prescribed Fermi level and temperature [50, 69] (cf. Section 4.1). The idea is that because of the very high number of electrons in the metal and the accompanying strong electron-electron scattering, the metal electrons are essentially in equilibrium. We thus assume the equilibrium distribution for the electrons in the metal

$$f_1(\mathbf{k}_1) = f^0\left(\frac{1}{k_B T}[W_1(\mathbf{k}_1) + W_{cM} - \phi_M]\right),\tag{6.27}$$

where $W_1(\mathbf{k}_1)$ is the energy dispersion of the metal conduction band, W_{cM} the conduction band edge in the metal, and ϕ_M the Fermi energy of the metal. The function f^0 is the Fermi-Dirac distribution

$$f^0(x) = \frac{1}{1 + \exp(x)}.\tag{6.28}$$

The temperature T has the same constant value for semiconductor electrons, metal electrons, semiconductor lattice, metal lattice, and environment, as is characteristic for the drift-diffusion transport model.

In (6.26), f_1 is required as a function of the wave vector $\mathbf{k}_2$ of the semiconductor, not of $\mathbf{k}_1$. Inserting the relation between $\mathbf{k}_1$ and $\mathbf{k}_2$ resulting from (4.5) into (6.27), we obtain with (6.2)

$$f_1^0(\mathbf{k}_2) \overset{!}{=} f_1(\mathbf{k}_1(\mathbf{k}_2)) = f^0\left(\frac{1}{k_B T}\left[\frac{\hbar^2 k_2^2}{2m_n} + W_{cS} - \phi_M\right]\right)\tag{6.29}$$

the necessary formulation, where we assumed a spherical parabolic band with effective mass m_n in the semiconductor.

The assumption for the distribution function in the *semiconductor* is somewhat more complicated. Since we consider drift-diffusion, the distribution in the volume of the semiconductor follows the description of Section 2.3. Hence, the ansatz (2.13) would be the obvious choice. However, the volume distribution becomes invalid near the interface, since the vicinity of the interface may cause a strong distortion of the distribution.

A typical distribution at the interface in non-equilibrium is characterized by a discontinuity at $v_{2x} = 0$. Figure 6.9 shows a typical appearance of a volume and an interface distribution. In equilibrium, the number of elec-

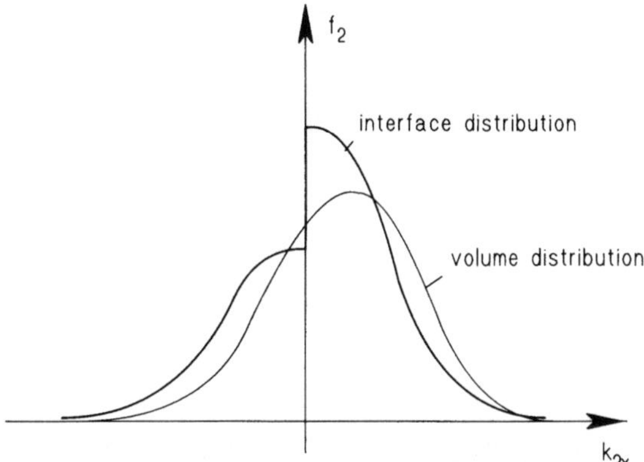

Fig. 6.9 Electron distributions in the volume and at the interface

trons entering the semiconductor from the metal is equal to the number of electrons leaving the semiconductor for the metal. Hence, in equilibrium no discontinuity appears. In non-equilibrium—say the chemical potential of the metal electrons is rised with respect to the semiconductor's—more electrons are entering the semiconductor from the metal than the semiconductor ist able to send back. This behaviour is displayed by the discontinuity in Fig. 6.9. A net current of electrons is then flowing from the metal to the semiconductor. Similarly, if the chemical potential of the metal electrons is lowered, fewer electrons are able to enter the semiconductor, and the outflowing electrons dominate the distribution function of the semiconductor. Also a discontinuity arises, but in this case with a reversed sign. The described discontinuity in the semiconductor distribution function near a material interface has been confirmed by Monte Carlo simulations [112]. Clearly, the discontinuity appears only strictly at the interface. The distribution undergoes a transition from the interface pattern to the volume form when going from the interface into the semiconductor volume. Since the reason for this transition are scattering processes which wash out the interface effects, the transition occurs roughly on the length of the mean free path of the semiconductor electrons. These qualitative statements on the transition region from the interface to the bulk distribution are confirmed by investigations of Baccarani [113], who analyzed the problem with the help of an approximate solution of the Boltzmann equation.

The above reflections suggest two possibilities for a self-consistent assumption of f_2:

1. a function describing a transition from the interface distribution to the volume distribution. This assumption would be the most accurate one because it can be matched to the interface distribution as well as the to the volume distribution. However, an additional model is necessary

    which describes the transition, since the boundary condition considered here determines the distribution only at the interface.

2. the volume distribution. This assumption matches the distribution in the interior of the semiconductor to the boundary condition. However, it has a lower accuracy since it neglects the exact shape of the distribution at the interface. It is a matching of the volume distribution on a coarser scale, ignoring the thin boundary layer in the vicinity of the interface.

Unfortunately, in the literature the most commonly used version is none of these possibilities, but merely a "crippled" version of the first one. It consists of

3. a function of the form of the interface distribution as in Fig. 6.9, but without the possibility of describing the transition to the bulk.

While this assumption is precise at the interface, the transition to the volume distribution is not included at all, thus bearing a certain level of inconsistency. The connection to the bulk is introduced only afterwards by an additional assumption, which will be given below.

Because of the widespread use of assumption 3, we are going to use it in the following despite its inherent inconsistency. In this way the model can be immediately compared to the very common Schottky barrier model of Crowell and Sze [79] as well as several other common models including some tunnel models [30, 104, 83, 106, 114]. An example for possibility 2, i.e. consideration of the volume distribution in the boundary condition, will be discussed in Section 6.3.

The assumption thus chosen is

$$f_2(\mathbf{k}_2) = \begin{cases} f_2^+(\mathbf{k}_2), & (k_{2x} \geq 0), \\ f_2^-(\mathbf{k}_2), & (k_{2x} < 0), \end{cases} \tag{6.30}$$

where f_2^+ describes the distribution of the incoming electrons, and f_2^- the distribution of the electrons leaving the semiconductor for the metal. If $f_2^+ \neq f_2^-$ at $k_{2x} = 0$, the discontinuity of Fig. 6.9 is reproduced. The functions f_2^+ and f_2^- are defined by

$$f_2^\pm(\mathbf{k}_2) \overset{!}{=} f^0\left(\frac{1}{k_B T}\left[\frac{\hbar^2 k_2^2}{2m_n} + W_{cS} - \phi_n^\pm\right]\right), \tag{6.31}$$

i.e. the equilibrium distribution, but with different quasi-Fermi energies for the inflow and outflow parts of the distribution. In thermodynamic equilibrium, ϕ_n^+ and ϕ_n^- are both equal to the overall equilibrium Fermi energy, and f_2 becomes the equilibrium distribution as required. In non-equilibrium, ϕ_n^+ and ϕ_n^- are different, and the asymmetry of f_2 results in a net current flow.

Now that the assumptions for f_1 and f_2 in (6.22) have been assembled, we can proceed by inserting (6.29) and (6.30) into the moment boundary condition (6.26), making use of the property that f_1^0 and f_2^- are even functions of k_{2x}. Hence we obtain as the metal-semiconductor contact boundary condition for the electron continuity equation the result

$$J_{nx} = -\frac{q}{4\pi^3} \int_0^\infty dk_{2x}\, v_{2x} D(k_{2x}) \int_{-\infty}^\infty \int_{-\infty}^\infty dk_{2y}\, dk_{2z}$$
$$\times\, [f_1^0(\mathbf{k}_2) - f_2^-(\mathbf{k}_2)]. \tag{6.32}$$

The boundary condition is in a form which expresses the electron current density J_{nx} as a function of the Fermi energies ϕ_M and ϕ_n^- of Eqs. (6.27) and (6.31).

The remaining question with (6.32) is: Where do we get ϕ_n^- from? The solver of the semiconductor devices equations determines the quasi-Fermi energy ϕ_n for the semiconductor volume, while ϕ_n^- is the equivalent quasi-Fermi level of the outflowing electrons at the interface. As discussed above, a transition region between the interface and the volume distribution exists. Knowledge of this transition would give us a relation between ϕ_n and ϕ_n^-. However, in the literature we are referring to, the common practice is to simply assume

$$\phi_n = \phi_n^- \tag{6.33}$$

as the required relation. Usually, no justifications are given for this assumption. It is inconsistent in the sense that it assumes a constant distribution for the outgoing electrons throughout the transition layer, while the distribution of the incoming electrons (characterized by ϕ_M) has to adapt to the bulk distribution. It is unreasonable to expect that only the incoming half of the distribution participates in the transition, while the other half remains unaffected. However, for the sake of comparability with the usual models, we keep assumption (6.33) for the moment.

6.2.3 Approximation of the Fermi-Dirac Distribution

The integrals in (6.32) cannot be evaluated analytically because of the occurence of the Fermi-Dirac distribution function

$$f^0(x) = \frac{1}{1 + e^x} \tag{6.34}$$

in f_1^0 and f_2^-. For device simulation purposes, however, a formula is required which can be calculated quickly in each iteration of the nonlinear solver. An approximation for f^0 becomes thus necessary.

The form of the Fermi-Dirac distribution function is shown in Fig. 6.10. The shape of f^0 suggests the idea of patching the function from appropriately determined exponential functions. The result of this consideration is

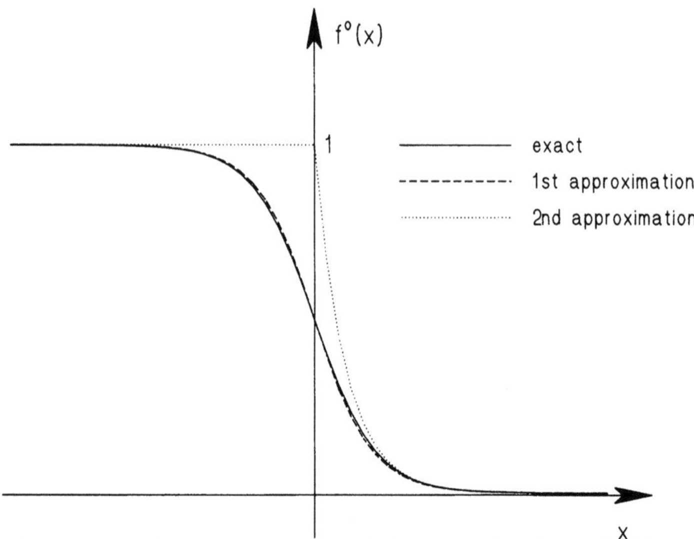

Fig. 6.10 Fermi-Dirac distribution function and approximations

$$f^0(x) \approx \begin{cases} 1 - e^x + \frac{1}{2}e^{3x/2} & \text{if } x < 0, \\ e^{-x} - \frac{1}{2}e^{-3x/2} & \text{if } x \geq 0, \end{cases} \tag{6.35}$$

where the constants appearing in the function have been selected such that the behaviour of the original f^0 is most satisfactorily reflected.

In detail, the behaviour of the approximation is in agreement with the original function with respect to the following points:

1. the asymptotic $f^0(x \to \infty) = e^{-x}$,
2. the asymptotic $f^0(x \to -\infty) = 1 - e^x$,
3. $f^0(0) = \frac{1}{2}$,
4. $df^0/dx|_{x=0} = -\frac{1}{4}$.

The above conditions force the shape of the approximation into that of the original function such that the remaining error is very small. The numerical constants in (6.35) have been determined in order to satisfy these conditions.

The graph of the approximation (6.35) is shown in Fig. 6.10 by the dashed line. As can be seen, (6.35) is indeed an excellent approximation of the Fermi-Dirac distribution.

While approximation (6.35) is sufficiently simple to allow an analytical evaluation of the contact current boundary condition (6.32), we make a further simplification in order to reduce the length of the formulas to follow. Eq. (6.35) is additionally simplified by keeping only the leading terms in (6.35):

$$f^0(x) \approx \begin{cases} 1 & \text{if} \quad x < 0 \quad (\textit{degeneration}), \\ e^{-x} & \text{if} \quad x \geq 0 \quad (\textit{non-degeneration}). \end{cases} \qquad (6.36)$$

The shape of (6.36) is represented in Fig. 6.10 by the dotted line. We can see that (6.36) still preserves the asymptotic behaviour of the Fermi-Dirac distribution while deviating from (6.34) in the vicinity of $x = 0$. The deviation of (6.36) from $f^0(W/k_B T)$ occurs only in a relatively small energy range of $\pm 3k_B T$, corresponding to a range of ± 0.075 eV at a temperature of 300 K. The following derivation of the contact boundary condition will be carried out using the approximation (6.36), thus keeping the formulas much more easy to view at a glance. Comparison of the results with those of the more exact formula (6.35) has shown that the additional terms in general have a negligible effect.

6.2.4 Evaluation of the Contact Current

Now all the necessary approximations for an analytical evaluation of (6.32) have been introduced. We begin the evaluation by first transforming (6.32) into a mathematically more tractable form.

Eq. (6.32) is composed of a difference of two terms, one involving f_1^0 and one involving f_2^-. These two terms can be interpreted as the current of electrons going from the metal to the semiconductor and vice versa, respectively. Since both terms are very similar in form, it is convenient to compute one of them first and conclude the second one by analogy. Thus we define J_{MS} as the current density of electrons flowing from metal to semiconductor by

$$J_{MS} = -\frac{q}{4\pi^3} \int_0^\infty dk_{2x}\, v_{2x} D(k_{2x}) \int_{-\infty}^\infty \int_{-\infty}^\infty dk_{2y}\, dk_{2z}\, f_1^0(\mathbf{k}_2). \qquad (6.37)$$

Remember that f_1^0 is the distribution function in the metal.

We are now going to transform (6.37) into a form which is accessible for easy analytical evaluation. First we consider the integration with respect to k_{2y} and k_{2z}, which constitutes the tangential part $\mathbf{k}_t$ of the wave vector. Noting that the f_1^0 in (6.29) does not depend on the direction of $\mathbf{k}_2$, we can turn to tangential polar coordinates by the variable transformation

$$\begin{aligned} k_{2y} &= k_t \cos \varphi, \\ k_{2z} &= k_t \sin \varphi, \end{aligned} \qquad (6.38)$$

introducing k_t and φ as the new integration variables. The integration with respect to φ can be carried out, giving

$$\int_{-\infty}^{\infty} \int_{-\infty}^{\infty} dk_{2y}\, dk_{2z}\, f^0 \left(\frac{1}{k_B T} \left[\frac{\hbar^2 k_2^2}{2m_n} + W_{cS} - \phi_M \right] \right)$$

$$= 2\pi \int_0^{\infty} dk_t\, k_t f^0 \left(\frac{k_t^2}{\alpha^2} + \frac{\phi_B - W_T}{k_B T} + \frac{k_{2x}^2}{\alpha^2} \right). \tag{6.39}$$

On the left-hand side we substituted f_1^0 from (6.29). On the right-hand side the abbreviation α has been introduced by

$$\frac{1}{\alpha^2} = \frac{\hbar^2}{2m_n k_B T}. \tag{6.40}$$

Furthermore, the difference $W_{cS} - \phi_M$ has been replaced by more convenient quantities using the relation

$$W_{cS} - \phi_M = \phi_B - W_T, \tag{6.41}$$

which follows from (6.8) and (6.9).
In the next step, we substitute the argument of f^0 in (6.39) as the new integration variable, defining

$$v_t = \frac{k_t^2}{\alpha^2} + \frac{k_{2x}^2}{\alpha^2} + \frac{\phi_B - W_T}{k_B T}. \tag{6.42}$$

We also introduce

$$v_x = \frac{k_{2x}^2}{\alpha^2}, \tag{6.43}$$

$$\eta_T = \frac{W_T}{k_B T}, \tag{6.44}$$

and

$$\psi_B = \frac{\phi_B}{k_B T} \tag{6.45}$$

for shorthand notation, and obtain for the right-hand side of (6.39)

$$2\pi \int_0^{\infty} dk_t\, k_t f^0 \left(\frac{k_t^2}{\alpha^2} + \frac{\phi_B - W_T}{k_B T} + \frac{k_{2x}^2}{\alpha^2} \right)$$

$$= \pi \alpha^2 \int_{v_x - \eta_T + \psi_B}^{\infty} dv_t\, f^0(v_t). \tag{6.46}$$

The v_t-integral in (6.46) is a function only of its lower bound. This enables us to define

$$F(v) \overset{!}{=} \int_v^{\infty} dv_t\, f^0(v_t), \tag{6.47}$$

allowing a very compact notation of the tangential wave vector integral in the subsequent formulas.

Next we turn to the k_{2x}-integration in (6.37). Since we have used spherical parabolic bands in the semiconductor distribution function (6.31), the group velocity v_{2x} according to (2.5) is given by

$$v_{2x} = \frac{\hbar k_{2x}}{m_n}. \tag{6.48}$$

Thus the current J_{MS} of (6.37) becomes with (6.46) and (6.47)

$$J_{MS} = -\frac{q}{4\pi^3} \frac{\hbar}{m_n} \pi\alpha^2 \int_0^\infty dk_{2x}\, k_{2x} D(k_{2x}) \cdot F(v_x(k_{2x}) - \eta_T + \psi_B). \tag{6.49}$$

We now substitute v_x from (6.43) for the integration variable, arriving at

$$J_{MS} = -\frac{q}{4\pi^3} \frac{\hbar}{m_n} \frac{\pi}{2} \alpha^4 \int_0^\infty dv_x\, D(\sqrt{\alpha}v_x) \cdot F(v_x - \eta_T + \psi_B). \tag{6.50}$$

The term $W_{2x} - W_0$, occurring in the expression for the transition probability (6.21), is rewritten into the new notation by inserting (6.43) and (6.8) into (6.20):

$$W_{2x} - W_0 = k_B T(v_x - \eta_T). \tag{6.51}$$

At this point we note that both factors of the integrand in (6.50) are functions of $v_x - \eta_T$. This enables us to simplify the equation by making a transformation of the integration variable to v_x^* with

$$v_x^* = v_x - \eta_T, \tag{6.52}$$

giving

$$\boxed{J_{MS} = -\frac{qm_n(k_B T)^2}{2\pi^2\hbar^3} \int_{-\eta_T}^\infty dv_x^*\, D(v_x^*) \cdot F(v_x^* + \psi_B)} \tag{6.53}$$

for the current J_{MS}. The prefactor of the integral has been condensed using (6.40). Note that η_T now enters the formula only as the lower bound of the integral. This will allow us to determine η_T resp. W_T as the energy range below the barrier peak where tunneling occurs (cf. Fig. 6.5 and the discussion at the beginning of Section 6.2).

The tunneling probability in (6.53) becomes in terms of v_x^* with (6.21), (6.51), and (6.52)

$$D(v_x^*) = \begin{cases} \exp(-\delta v_x^{*2}) & \text{if} \quad -\eta_T \leq v_x^* \leq 0 \quad \text{(tunneling)}, \\ 1 & \text{if} \quad 0 < v_x^* < \infty \quad \text{(therm. emission)}, \end{cases} \tag{6.54}$$

where

$$\delta = \frac{(k_B T)^2}{q V_D W_{00}} \tag{6.55}$$

has been introduced as an abbreviation. Note that $v_x^* < 0$ corresponds to tunneling, while $v_x^* > 0$ corresponds to thermionic emission. Eq. (6.53) together with (6.47) and (6.54) represents a manageable set of equations for the following final evaluation of the contact current.

Now we shall evaluate (6.53). In the first place, the function F, defined in (6.47), has to be specified. At this point, the approximation of subsection 6.2.3 comes into play. If we want to use the second approximation, we obtain by inserting (6.36) into (6.47)

$$F(y) = \begin{cases} 1 - y & \text{if } y \le 0 \quad \text{(degeneration)}, \\ \exp(-y) & \text{if } y > 0 \quad \text{(non-degeneration)}. \end{cases} \tag{6.56}$$

Otherwise, if the more precise approximation (6.35) is desired, insertion of (6.35) into (6.47) gives the more exact but also more complicated formulation.

Now (6.56) has to be substituted into (6.53). Here we have to distinguish the two cases where the point $y = 0$ of (6.56) resp. $v_x^* = -\psi_B$ of (6.53) is inside or outside the integration interval in (6.53). This distinction corresponds to the cases $\psi_B < \eta_T$ or $\psi_B > \eta_T$, respectively. If the barrier height ϕ_B in Fig. 6.5 is smaller than the tunneling energy range W_T (i.e. $\psi_B < \eta_T$), the physical meaning is that in equilibrium the semiconductor at the boundary is in degeneration, because the Fermi energy lies above $W_c(x_T)$. Alternatively, in equilibrium the semiconductor is non-degenerate if $\psi_B > \eta_T$ is valid. Starting with the degeneration case, insertion of (6.56) into (6.53) yields

$$
\begin{aligned}
J_{MS} = &-\frac{q m_n (k_B T)^2}{2\pi^2 \hbar^3} \int_{-\eta_T}^{-\psi_B} dv_x^* \, D(v_x^*) \cdot [1 - v_x^* - \psi_B] \\
&-\frac{q m_n (k_B T)^2}{2\pi^2 \hbar^3} \int_{-\psi_B}^{\infty} dv_x^* \, D(v_x^*) \cdot \exp(-v_x^* - \psi_B).
\end{aligned}
\tag{6.57}
$$

Note how the factor of $D(v_x^*)$ in the integrand switches from linear to exponential at $v_x^* = -\psi_B$ according to (6.56). In the non-degeneration case, $v_x^* = -\psi_B$ lies outside the integration interval, and only the exponential branch of (6.56) enters into the integral:

$$J_{MS} = -\frac{q m_n (k_B T)^2}{2\pi^2 \hbar^3} \int_{-\eta_T}^{\infty} dv_x^* \, D(v_x^*) \cdot \exp(-v_x^* - \psi_B). \tag{6.58}$$

Next, the analytical form of the transition probability $D(v_x^*)$ of (6.54) has to be introduced into (6.57) and (6.58). According to the two branches in (6.54), the integration intervals in (6.57) and (6.58) have to be split at $v_x^* = 0$.

The result of inserting (6.54) in the degenerate case (6.57) is

$$J_{MS} = -\frac{qm_n(k_BT)^2}{2\pi^2\hbar^3} \int_{-\eta_T}^{-\psi_B} dv_x^* \, \exp(-\delta v_x^{*2}) \cdot [1 - v_x^* - \psi_B]$$

$$-\frac{qm_n(k_BT)^2}{2\pi^2\hbar^3} \int_{-\psi_B}^{0} dv_x^* \, \exp(-\delta v_x^{*2}) \cdot \exp(-v_x^* - \psi_B)$$

$$-\frac{qm_n(k_BT)^2}{2\pi^2\hbar^3} \int_{0}^{\infty} dv_x^* \, \exp(-v_x^* - \psi_B), \tag{6.59}$$

where we presupposed $\psi_B > 0$. It should be noted here that the quasi-Fermi level of electrons travelling towards the metal may come to lie above the barrier peak under strong forward bias and strong tunneling, which corresponds to $\psi_B < 0$. In this case it is the first integral of (6.57) that has to be splitted, not the second as in (6.59).

In the non-degenerate case (6.58) we obtain

$$J_{MS} = -\frac{qm_n(k_BT)^2}{2\pi^2\hbar^3} \int_{-\psi_B}^{0} dv_x^* \, \exp(-\delta v_x^{*2}) \cdot \exp(-v_x^* - \psi_B)$$

$$-\frac{qm_n(k_BT)^2}{2\pi^2\hbar^3} \int_{0}^{\infty} dv_x^* \, \exp(-v_x^* - \psi_B). \tag{6.60}$$

Examination of (6.59) and (6.60) reveals that besides the simple integral of the exponential function the integrals $\int dx \, e^{-\delta x^2}$, $\int dx \, xe^{-\delta x^2}$, and $\int dx \, xe^{-\delta x^2 - x}$, are occurring. Evaluation of these integrals as indefinite integrals yields:

$$\int dx \, \exp(-\delta x^2) = \frac{1}{2}\sqrt{\frac{\pi}{\delta}} \, \text{erf}(\sqrt{\delta}x), \tag{6.61}$$

$$\int dx \, x \exp(-\delta x^2) = -\frac{1}{2\delta} \exp(-\delta x^2), \tag{6.62}$$

$$\int dx \, \exp(-\delta x^2 - x) = \frac{1}{2}\sqrt{\frac{\pi}{\delta}} \exp\left(\frac{1}{4\delta}\right) \text{erf}\left(\sqrt{\delta}x + \frac{1}{2\sqrt{\delta}}\right). \tag{6.63}$$

Making use of (6.61), (6.62), and (6.63) in (6.59) and (6.60) gives the final result, shown in Table 6.2. Eqs. (6.64) resp. (6.65) represent the current density of electrons flowing from the metal to the semiconductor in the case of degenerate resp. non-degenerate semiconductor. Note that the first four terms in (6.64) and the first two terms in (6.65), respectively, represent the tunneling current, while the last term in each case represents the thermionic emission part of the current.

This concludes the evaluation of the first part of (6.32). Next, we have to compute the other term in (6.32) which involves f_2^-. This term can be

Table 6.2. *Contact current densities*

$$
\begin{aligned}
J_{MS} = & -\frac{qm_n(k_BT)^2}{2\pi^2\hbar^3}\frac{1}{2}\sqrt{\frac{\pi}{\delta}}\left\{\exp\left(-\psi_B + \frac{1}{4\delta}\right)\right. \\
& \times\left[\mathrm{erf}\left(\frac{1}{2\sqrt{\delta}}\right) - \mathrm{erf}\left(\frac{1}{2\sqrt{\delta}} - \sqrt{\delta}\psi_B\right)\right] \\
& + (\psi_B - 1)[\mathrm{erf}(\sqrt{\delta}\psi_B) - \mathrm{erf}(\sqrt{\delta}\eta_T)] \\
& \left. + \frac{1}{\sqrt{\pi\delta}}[\exp(-\delta\psi_B^2) - \exp(-\delta\eta_T^2)]\right\} \\
& -\frac{qm_n(k_BT)^2}{2\pi^2\hbar^3}\exp(-\psi_B) \qquad \text{if } (\psi_B < \eta_T) \qquad (6.64)
\end{aligned}
$$

$$
\begin{aligned}
J_{MS} = & -\frac{qm_n(k_BT)^2}{2\pi^2\hbar^3}\frac{1}{2}\sqrt{\frac{\pi}{\delta}}\exp\left(-\psi_B + \frac{1}{4\delta}\right) \\
& \times\left[\mathrm{erf}\left(\frac{1}{2\sqrt{\delta}}\right) - \mathrm{erf}\left(\frac{1}{2\sqrt{\delta}} - \sqrt{\delta}\eta_T\right)\right] \\
& -\frac{qm_n(k_BT)^2}{2\pi^2\hbar^3}\exp(-\psi_B) \qquad \text{if } (\psi_B > \eta_T) \qquad (6.65)
\end{aligned}
$$

interpreted as the current density of electrons flowing from the semiconductor to the metal; it is defined by

$$
J_{SM} = -\frac{q}{4\pi^3}\int_0^\infty dk_{2x}\, v_{2x}D(k_{2x})\int_{-\infty}^\infty\int_{-\infty}^\infty dk_{2y}\, dk_{2z}\, f_2^-(\mathbf{k}_2). \quad (6.66)
$$

As stated in the beginning of this subsection, we are going to infer J_{SM} from J_{MS} by analogy. Eq. (6.66) evolves from (6.37) by substitution of f_1^0 with f_2^-. Referring to (6.29) and (6.31), we note that f_2^- emerges if ϕ_M is replaced by ϕ_n^-:

$$
f_1^0 \to f_2^- \quad \Leftrightarrow \quad \phi_M \to \phi_n^-. \qquad (6.67)
$$

With the assumption $\phi_n = \phi_n^-$, which was discussed in Subsection 6.2.2 in conjunction with Eq. (6.33), this means that we have to make the substitution

$$
\phi_M \to \phi_M + qV_c, \qquad (6.68)
$$

where we made use of the definition (6.6) of the contact voltage drop V_c. From (6.9) follows that the barrier height, in turn, has to be replaced according to

$$
\phi_B = W_0 - \phi_M \quad \to \quad W_0 - \phi_M - qV_c = \phi_B - qV_c. \qquad (6.69)
$$

As a result, J_{SM} is obtained from the formula for J_{MS} by substituting ψ_B in

(6.64) and (6.65) with

$$\psi_B \rightarrow \psi_B - qV_c/k_B T.$$
(6.70)

The total contact current density is then given by

$$J_{nx} = J_{MS} - J_{SM}.$$
(6.71)

We relinquish here to write down the full expression for J_{SM}, since it is fully identical with (6.64) and (6.65) except for the substitution (6.70).

Before we will discuss the result of this subsection, we are going to complete the metal-semiconductor contact boundary condition by deriving the boundary condition for the Poisson equation in the following subsection.

6.2.5 Boundary Condition for the Poisson Equation

The solution of the Poisson equation is the distribution of the electrostatic potential φ. Thus, in order to derive a boundary condition for the Poisson equation, we have to look for a condition which determines φ at the contact boundary in consistence with the preceding model concept. Since the potential φ and the semiconductor conduction band edge W_c differ only by the electron affinity χ, a corresponding condition for W_c is equivalent.

It has been stated in the introduction of Section 6.2 that the actual boundary x_T for the simulation domain is placed at the edge of the tunneling region. From Fig. 6.5 we find for the band edge at x_T

$$W_c(x_T) = \phi_M + \phi_B - W_T,$$
(6.72)

where W_T is the tunneling energy range, and ϕ_M is the Fermi energy in the metal. Since we define the zero energy as the Fermi energy in equilibrium, ϕ_M is directly proportional to the voltage applied to the contact.

The condition (6.72) can be regarded as an apparent reduction of barrier height due to the tunneling effect. However, this should not be confused with simpler models where the tunneling current is modelled by thermionic emission over a reduced barrier [92, 30].

While (6.72) is already the desired boundary condition for the Poisson equation, we are still left with the determination of W_T resp. x_T. According to the discussion at the beginning of Section 6.2, W_T will be fixed by the condition that tunneling can be neglected for $x > x_T$. This means that on one hand W_T must be made as small as possible, in order that the boundary x_T is closest to the true metallurgical junction. On the other hand, W_T must be as large as necessary so that the tunneling process occurs predominantly in this energy range.

Recall that from (6.53) the current density of electrons flowing from metal to semiconductor can be written as

$$J_{MS} = -\frac{qm_n(k_B T)^2}{2\pi^2\hbar^3} \int_{-\eta_T}^{\infty} dv_x^* \, D(v_x^*) \cdot F(v_x^* + \psi_B). \tag{6.73}$$

The integral has already been set up in Subsection 6.2.4 such that the desired quantity W_T occurs only at a single place, namely at the lower bound of the integral. This enables us to determine η_T by the condition that the dominant part of the tunneling is included in the integration interval, the rest being negligible.

According to (6.54), the interval $[-\eta_T, 0]$ describes tunneling, while $(0, \infty)$ corresponds to thermionic emission. With decreasing $v_x^* < 0$, the transition probability rapidly approaches zero. The lower bound of the integral now must be chosen such that the dominant contribution to the integral is contained in the integration interval, and further increase of $|\eta_T|$ contributes only negligibly to the integral.

We proceed as follows. The diagram in Fig. 6.11 schematically shows the dependency of the integrand from the integration variable. First, we determine the maximum of the integrand $D \cdot F$. The value of the integration variable at the maximum of the integrand is labeled v_{xm}^*. Then $-\eta_T$ is placed where the integrand becomes much smaller than its value at the maximum (cf. Fig. 6.11):

$$D(\eta_T)F(\eta_T + \psi_B) \ll D(v_{xm}^*)F(v_{xm}^* + \psi_B). \tag{6.74}$$

Since the integrand drops to zero exponentially, the dominant contribution to the integral is contained in the integration interval if $|\eta_T|$ is fixed at the smallest value that satisfies (6.74). To be specific, we choose $-\eta_T$ at that point where the integrand has dropped to the e^{-3}-fold of its maximum, i.e.

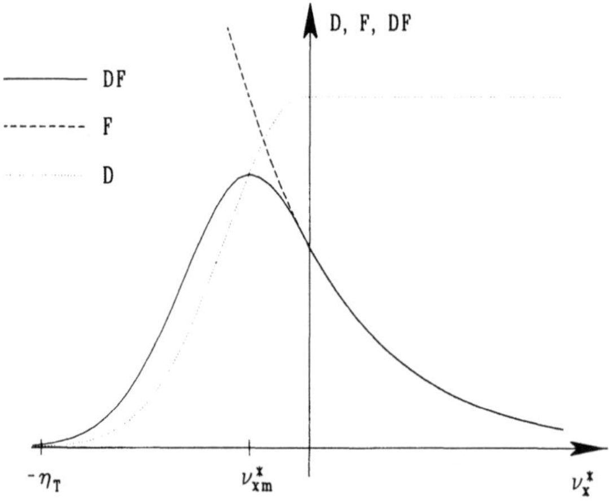

Fig. 6.11 Determination of η_T

we *define* the meaning of "much smaller" by setting

$$D(\eta_T)F(\eta_T + \psi_B) = \exp(-3)\cdot D(v^*_{xm})F(v^*_{xm} + \psi_B). \tag{6.75}$$

Since $\exp(-3)$ is approximately 0.05, this seems to be an acceptable choice. The foregoing program is carried out in detail in Appendix C. As a result we find

$$\eta_T = \begin{cases} \sqrt{v^{*2}_{xm} + \dfrac{3}{\delta}}, & \psi_B < \eta_T, \psi_B < \dfrac{1}{2\delta}, \\[2ex] \sqrt{\dfrac{\psi_B + 3}{\delta} - \dfrac{1}{4\delta^2}}, & \psi_B < \eta_T, \psi_B \geq \dfrac{1}{2\delta}, \\[2ex] \dfrac{1}{2\delta} + \sqrt{\dfrac{3}{\delta}}, & \psi_B \geq \eta_T, \end{cases} \tag{6.76}$$

where v^*_{xm} is defined by

$$v^*_{xm} = -\frac{1}{2}\left[\psi_B - 1 + \sqrt{(\psi_B - 1)^2 + \frac{2}{\delta}}\right]. \tag{6.77}$$

Eq. (6.76) determines η_T and hence W_T as a function of the barrier height and effective doping. With W_T known, the boundary condition for the Poisson equation (6.72) is fully determined.

However, recall that we computed η_T from (6.53), which is only a part of the total current, namely the current J_{MS} of electrons going from the metal to the semiconductor. According to (6.71), the total current is computed from the superposition of electrons travelling to the right as well as to the left. It is clear that η_T has to be determined such that the tunneling currents of both directions are fully accounted for. Thus we have to compute actually two η_T's, one from J_{MS} and one from J_{SM}. The smaller η_T is contained in the larger one, so selecting the larger of the two ensures that both tunneling currents are fully captured by the tunneling energy range.

The basic behaviour of η_T is depicted in Fig. 6.12. The figure shows the apparent barrier lowering η_T as a function of δ for a typical range of values. The value of ψ_B has been set to 30 in Fig. 6.12, corresponding to a barrier height of about 0.75 eV at a temperature of 300 K. Note that by Eqs. (6.55) and (6.14), δ decreases if N_D^* increases; correspondingly N_D^* rises as we go along the horizontal axis in Fig. 6.12 from right to left. For very high values of δ, η_T is nearly zero. Tunneling is negligible there, and we are in the regime of thermionic emission. As N_D^* becomes larger, η_T rises, indicating the onset of tunneling in the thermionic field emission regime. A kink in the curve appears when the semiconductor turns from non-degeneration to the degenerate case. In the subsequent field emission regime, η_T rises sharply again with increasing N_D^*, indicating that tunneling now becomes the dominant transport process.

An important feature of Fig. 6.12 is the strong increase of η_T with decreasing

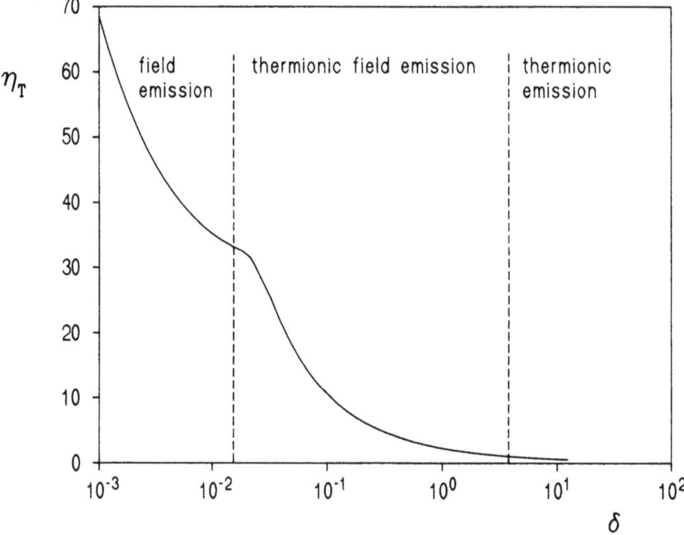

Fig. 6.12 Barrier lowering η_T as a function of δ

δ. This means that the tunneling energy range W_T quickly becomes large with increasing N_D^*, as is clear from the discussion at the beginning of Section 6.2. However, W_T cannot grow infinitely, but only as long as the preconditions of the present model are not violated. These were:

1. The assumed parabolic behaviour of the conduction band edge (6.10) limits the consideration to the depletion region.
2. Tunneling through the barrier according to (6.12) makes only sense from points on the flank of the barrier. This limits the consideration to the left half of the parabola (6.10).

The consequence of the first condition is demonstrated in Fig. 6.13. It shows the bending of the conduction band edge of a one-dimensional contact in a reversed bias condition (to an exaggerated degree). The band edge is curved in the depletion region, while in the quasi-neutral zone it becomes a declining (more or less) straight line. In any case, due to the assumption (6.10) of a parabolic behaviour of the conduction band edge, the result for η_T (6.76) is only valid as long as $W_0 - W_T$ is still in the depletion region. As soon as W_T becomes so large that x_T reaches the quasi-neutral zone, we actually should determine W_T from a similar consideration of the linearly declining band edge. However, in order to keep the model simple and manageable, we neglect instead the contribution of the tunneling from states in the interior of the neutral zone by *pinning x_T at the edge of the depletion region*. As a boundary condition for the Poisson equation, (6.72) is then replaced by the neutrality condition

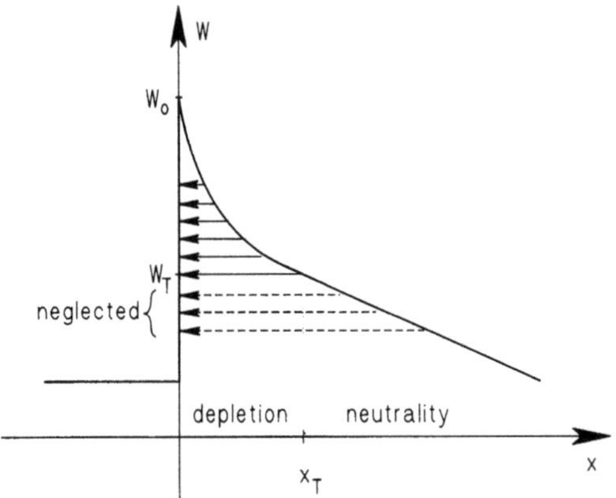

Fig. 6.13 Tunneling range in reverse condition

$$n - p - N_D + N_A = 0. \tag{6.78}$$

From this equation, W_c at the boundary can be determined. Inserting (2.22) and (2.24) into (6.78), we obtain

$$\gamma_n N_c \exp\left(\frac{\phi_n - W_c}{k_B T}\right) - \gamma_p N_v \exp\left(\frac{W_c - W_g - \phi_p}{k_B T}\right)$$

$$= N_D - N_A, \tag{6.79}$$

where W_v has been expressed by W_c and the bandgap W_g using

$$W_v = W_c - W_g. \tag{6.80}$$

Solving (6.79) for W_c yields the Poisson equation boundary condition directly for the conduction band edge:

$$\boxed{\begin{aligned} W_c &= \frac{1}{2}(\phi_n + \phi_p + W_g) + \frac{k_B T}{2} \ln\left(\frac{N_c \gamma_n}{N_v \gamma_p}\right) \\ &\quad - k_B T \operatorname{Arsinh}\left(\frac{N_D - N_A}{2\sqrt{\gamma_n \gamma_p N_c N_v}} \exp\left[\frac{\phi_p - \phi_n + W_g}{2 k_B T}\right]\right). \end{aligned}}$$

$$\tag{6.81}$$

In this case, W_T is no longer given by (6.76). Instead, (6.76) is replaced by (6.72), solved for W_T,

$$W_T = \phi_B + \phi_M - W_c(x_T), \tag{6.82}$$

because now $W_c(x_T)$ is known from (6.81).

Eq. (6.81) can be written more compact (see (6.149) below) by introduction of the "effective intrinsic concentration" n_{ie} [21], which is defined by

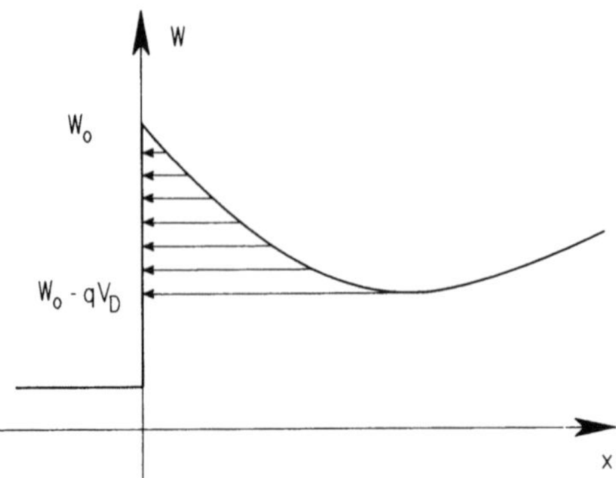

Fig. 6.14 Tunneling range in forward condition

$$n_{ie} = \sqrt{\gamma_n \gamma_p N_c N_v}\, \exp\left(-\frac{W_g}{2k_B T}\right). \tag{6.83}$$

The effective intrinsic concentration n_{ie} is a convenient parameter accounting for bandgap narrowing in the heavy doping case [21]. It can be approximated by its value in equilibrium, which depends only on the doping concentration for a given semiconductor [115, 116].

The circumstances of the second precondition above are demonstrated in
Fig. 6.14, showing the contact in a forward bias condition. Again, the band
edge is curved in the depletion region, but now it goes through a minimum
before reaching the quasi-neutral zone, where it becomes a rising (more or
less) straight line. At the minimum, W_T is equal to qV_D. As indicated in the
figure, tunneling through the barrier occurs only from positions on the
falling slope of the barrier. In this case, x_T has to be pinned at the band edge
minimum, corresponding to $E_x(x_T) = 0$ and $W_T = qV_D$. This condition will
be observed during the determination of V_D in the following subsection by
restricting $E_x(x_T)$ to negative values.

In summary we note that the boundary condition for the Poisson equations
consists of either (6.72) or (6.78), depending on the value of W_T. This distinction arises because the formula (6.76) for η_T looses its validity as soon as x_T
reaches the edge of the depletion region. This is the case when the depletion
W_T from (6.76) becomes larger than the neutrality W_T from (6.82). The
model then switches from (6.72) and (6.76) to (6.78) and (6.82).

6.2.6 Computation of the Maximum Field

In the previous subsection, we concluded the derivation of the contact
model by deriving formulas for η_T as a function of δ and ψ_B. In a practical
implementation, however, an additional complication arises since some

necessary quantities are not explicitly known by a simulator. We shall address this problem in the present subsection.

The quantity δ occurring in the formulas for η_T has been introduced as an abbreviation in Eq. (6.55). The definition contains V_D, which in turn has been defined in (6.11) as

$$qV_D = \frac{\varepsilon_s E_{\max}^2}{2N_D^*}.$$

(6.84)

As indicated in several of the figures on the preceding pages, V_D corresponds roughly to the diffusion potential of the contact's depletion region. However, the diffusion potential is the potential difference between the ends of the depletion region. Thus it is a non-local concept, which is important in an approximative analytical solution of the potential in a depletion region. A device simulator, instead, computes the potential by solving a locally discretized form of the transport equations; the non-local diffusion potential is not meaningful in this context. Thus, we consider (6.84) only as a formal definition of a quantity called V_D; it has the property that under certain idealizing conditions it is an estimate for the diffusion potential.

The definition (6.84) contains the two quantities $E_{\max}$ and N_D^* which require special attention. N_D^* has been introduced in subsection 6.2.1 as a parameter describing the curvature of the bands in the x-direction. For convenience, qN_D^* has been defined loosely speaking as the effective space charge responsible for the band bending in the x-direction. Here we need a more precise definition:

$$\frac{q^2 N_D^*}{\varepsilon_s} = \frac{\partial^2 W_c}{\partial x^2} \approx \text{const.}$$

(6.85)

The assumption of a parabolic barrier shape (6.10) implies that N_D^* is taken as approximately constant within the tunneling length. For implementation purposes, it is convenient to rewrite (6.85) with the help of the Poisson equation (2.29) as

$$N_D^* = \frac{1}{q}\rho_d - \frac{\varepsilon_s}{q^2}\left[\frac{\partial^2 W_c}{\partial y^2} + \frac{\partial^2 W_c}{\partial z^2}\right],$$

(6.86)

where ρ_d is the total space charge density in the depletion region. If x_T is inside the depletion region, the simulator can immediately calculate ρ_d from the charge concentrations on the boundary. In the case that the depletion region is fully outside of the simulation domain, we have charge neutrality on the boundary according to (6.78), and the simulator does not know the space charge in the depletion region. However, this is the case at very high doping concentrations, where the space charge of the minority carriers can be neglected. The space charge density in the depletion region then can be approximated by $q(N_D - N_A)$. This effect can simply be incorporated in (6.86) by computing ρ_d from

$$\rho_d = \begin{cases} q(N_D - N_A), & \text{if } q|N_D - N_A| > |\rho|, \\ \rho(x_T), & \text{otherwise.} \end{cases} \tag{6.87}$$

The second quantity of (6.84) that needs consideration is $E_{\max}$. It has been introduced in (6.10) as the maximum electric field in the barrier. However, according to Fig. 6.5, we placed the boundary of the simulation domain at $x = x_T$, while the maximum field occurs well inside the tunneling region, very near to the metallurgical junction. But this point is outside of the simulation domain, and the simulator does not explicitly know the electric field at this point. Therefore, $E_{\max}$ has to be extrapolated from the electric field on the boundary at x_T.

It follows from the parabolic band shape (6.10) that the electric field varies linearly over the tunneling length. We differentiate (6.10) with respect to x, set $x = x_T$, and solve for $E_{\max}$. This manipulation gives

$$E_{\max} = E_x(x_T) - \frac{qN_D^*}{\varepsilon_s} x_T, \tag{6.88}$$

where

$$E_x(x_T) = \frac{1}{q} \frac{\partial W_c}{\partial x}\Big|_{x=x_T} \tag{6.89}$$

has been introduced. According to the discussion in the previous subsection, $E_x(x_T)$ has to be restricted to negative values (if the contact is on n-conducting material). If the simulator should find a positive E_x on the boundary during the iteration, E_x has to be set to zero in (6.88) and the following equations in order to force the boundary onto the band edge minimum.

The tunneling length x_T in (6.88) is determined by setting $x = x_T$ in (6.10) and solving for x_T,

$$x_T = -\frac{\varepsilon_s E_{\max}}{qN_D^*}\left[1 - \sqrt{1 - \frac{W_T}{qV_D}}\right], \tag{6.90}$$

where use has been made of (6.84) and (6.8). Insertion of (6.90) into (6.88) and solving for $E_{\max}$ yields

$$E_{\max} = \frac{E_x(x_T)}{\sqrt{1 - \dfrac{W_T}{qV_D}}} \tag{6.91}$$

as the desired extrapolation formula for $E_{\max}$.

With the help of (6.91), we are able to eliminate $E_{\max}$ from (6.84), arriving at

$$qV_D - W_T(V_D) = \frac{\varepsilon_s}{2N_D^*} E_x^2(x_T). \tag{6.92}$$

We noticed explicitly in (6.92) that W_T depends on V_D through the formula (6.76) for η_T. Thus, (6.92) represents a nonlinear equation for V_D, which does not contain any more unknown quantities. Note that by (6.92), W_T can never exceed qV_D, as required in the discussion of Subsection 6.2.5.

Using the definition for δ (6.55), we can replace V_D with δ. Making this substitution in (6.92), and recalling that $\eta_T = W_T/(k_B T)$, we alternatively obtain a direct equation for δ:

$$\frac{1}{\delta} - \frac{W_{00}}{k_B T}\eta_T(\delta) = \frac{\varepsilon_s W_{00}}{2N_D^*(k_B T)^2} E_x^2(x_T). \tag{6.93}$$

It is suggestive to compute δ in the first place, because this is the tunneling parameter occurring in the current density expressions (6.64)–(6.65).

Eqs. (6.93) and (6.76) together are two equations for the two unknowns η_T and δ. They have to be solved simultaneously. This can be done conveniently by eliminating one variable in one equation with the help of the second equation; the resulting equation can be solved either directly or with the help of a Newton iteration. Here, we content ourselves with a discussion of the qualitative behaviour. Fig. 6.15 shows both equations in the form η_T versus δ (cf. also Fig. 6.12). Eq. (6.76) is represented by the solid line. Eq. (6.93) has been transformed for representation in the diagram into

$$\eta_T(\delta) = \frac{k_B T}{W_{00}} \frac{1}{\delta} - \left(\frac{E_x(x_T)}{E_0}\right)^2, \tag{6.94}$$

where E_0 stands for

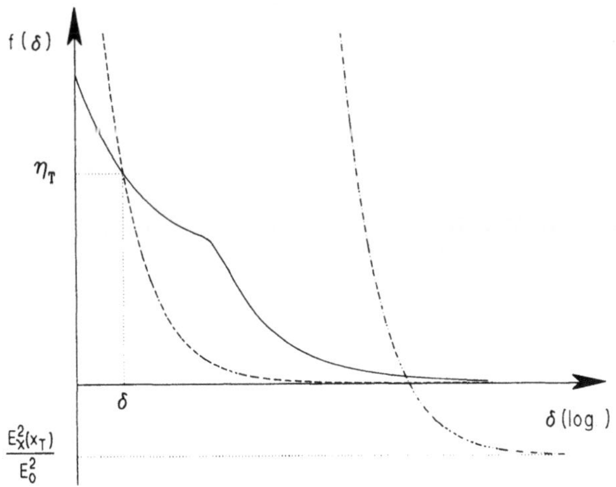

Fig. 6.15 Simultaneous determination of η_T and δ

$$E_0^2 = \frac{2N_D^*(k_B T)^2}{\varepsilon_s W_{00}}.$$ (6.95)

It is represented for the cases of high and low N_D^* by the dashed and the dash-dotted lines, respectively. The crossing point of the two graphs determines the simultaneous solution for η_T and δ.

The asymptotic behaviour of (6.76) for small δ is

$$\lim_{\delta \to 0} \eta_T = \sqrt{\frac{7}{2\delta}},$$ (6.96)

while for large δ it is

$$\lim_{\delta \to \infty} \eta_T = \sqrt{\frac{3}{\delta}}.$$ (6.97)

Thus, the overall behaviour of this function is $\eta_T \sim 1/\sqrt{\delta}$. If we compare this with the second curve (6.94), noticing that $1/\delta$ varies faster than $1/\sqrt{\delta}$, we can infer that an intersection must occur in any case.

If N_D^* is low, W_{00} is also small, and the $1/\delta$ curve is multiplied by a large factor in (6.94), giving the dash-dotted line in Fig. 6.15. Hence, the crossing point occurs at large δ and small η_T:

$$\lim_{N_D^* \to 0} W_T = 0.$$ (6.98)

This solution corresponds to vanishing tunnel current when the band bending is small.

If N_D^* becomes large on the other hand, the $1/\delta$-curve is always very near to the axes (dashed line), and the intersection occurs at small δ and large η_T. This solution corresponds to the strong tunneling regime. Note that in this limit the constant term in (6.94) and also in (6.92) becomes negligible. Under this condition, the equations can be solved analytically. The solution for $\eta_T(\delta)$ then becomes $k_B T/(W_{00}\delta)$; W_T and qV_D—according to (6.92)— converge to the same value:

$$\lim_{N_D^* \to \infty} W_T = qV_D.$$ (6.99)

This relation indicates that, as expected, for high doping the tunneling region encompasses the whole depletion region.

6.2.7 Minority Carriers and Bipolar Model

In the preceding sections, we considered only majority carriers, since these are the dominating charge transporting particles in our contact model. However, in order to be useful for device simulation, a consistent description of the transport behaviour of the *minority carriers* is required, too.

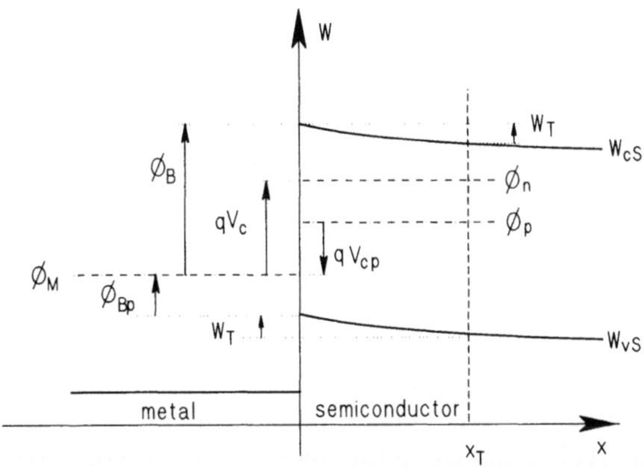

Fig. 6.16 Conduction and valence band at the contact

Because the device simulator regularly solves continuity equations for both electrons and holes, simultaneous boundary conditions at the contact for both equations are necessary.

The basic situation is depicted in Fig. 6.16. The figure shows the positional dependence of the semiconductor conduction and valence bands near the contact. As before, we consider the case of a contact to n-conducting material (cf. Fig. 6.5). We are already familiar with the behaviour of the conduction band, decreasing from the junction towards the bulk semiconductor and thus forming a barrier with respect to the electron flow. The barrier height ϕ_B, tunneling length x_T, voltage drop V_c, and tunneling energy range W_T are also familiar quantities which can be found again in the figure.

Clearly, the image of the valence band edge is a true copy of the conduction band, shifted downwards on the energy scale by the value of the semiconductor bandgap W_g. As discussed briefly in Subsection 6.1.1, the behaviour of the holes is quite different compared to the electrons. While the electrons see a potential barrier in their way which they either have to overcome or to tunnel through, the holes feel a force dragging them towards the junction because of the slope of the valence band. In contrast to the electrons, there is no potential *peak* where the holes could tunnel through. Thus, tunneling does not play a role in the transport of the minority carriers. Instead, the holes find a "pocket" at the interface where they tend to accumulate.

However, the holes encounter a potential *step* in their way when they cross the interface from the metal to the semiconductor. In the usual picture, a hole that travels in the semiconductor towards the contact finally reaches the metal and annihilates with a metallic electron which has approached the interface from the opposite side. Alternatively, we may consider the empty electron states in the metal as holes [30]. In this picture, all the states more than a few $k_B T$ above the Fermi level are "occupied by holes", while the number of holes decreases exponentially below the Fermi level. Rising the temperature excites electrons from below the Fermi energy to energies above, leaving empty states and thus increasing the number of holes below the Fermi level. A hole sufficiently deep in the metal band is able to cross the interface and to continue its way in the semiconductor valence band. In fact, this process can be regarded as *thermionic emission of holes.*

In summary, we find—with respect to thermionic emission—a symmetric situation for electrons and holes. In order to cross the interface, the electrons must have enough kinetic energy to overcome the barrier of height ϕ_B. Since the kinetic energy of the holes is measured downwards in Fig. 6.16, the holes in analogy must be deep in the metallic band to overcome the barrier, which for holes is between the semiconductor *valence* band and the metal Fermi level. Thus we define in Fig. 6.16 the hole barrier height ϕ_{Bp} by

$$\phi_{Bp} = \phi_M - W_{vS}, \tag{6.100}$$

where W_{vS} is the valence band edge of the semiconductor at the interface. With the semiconductor band gap energy

$$W_g = W_{cS} - W_{vS} \tag{6.101}$$

and (6.1), the hole barrier height can be expressed by the electron barrier height:

$$\phi_{Bp} = W_g - \phi_B. \tag{6.102}$$

From these relations follows immediately the normalized hole barrier height

$$\psi_{Bp} = \frac{\phi_{Bp}}{k_B T} = \frac{W_g - \phi_B}{k_B T} \tag{6.103}$$

as the equivalent of ψ_B. In correspondence to V_c, we also define in Fig. 6.16 the hole voltage drop as the difference between the metal Fermi level and the hole quasi-Fermi level,

$$qV_{cp} = \phi_M - \phi_p \tag{6.104}$$

since in non-equilibrium, ϕ_n and ϕ_p do not necessarily coincide.

Now all necessary ingredients have been collected in order to treat hole and electron thermionic emission equivalently on the same footing. However, recall that we formulated the boundary condition for the electrons at the position x_T; thus the minority carrier boundary condition also has to be specified at this point. The value of x_T has been determined as the length over which tunneling of electrons occurs. The tunneling length is very small; in the foregoing we neglected scattering processes in this region. Consequently, we also have to neglect scattering of holes within the tunneling region. Thus the holes undergo ballistic transport until they reach the boundary x_T, where the regime of the bulk transport equations begins.

According to Fig. 6.16, the total hole barrier height across the interface is $\phi_{Bp} + W_T$, because the valence band edge decreases by an additional amount of W_T over the length x_T.

The thermionic emission current of electrons going from the metal to the semiconductor has been given by the last term in any case of (6.64) or (6.65), respectively. By analogy, we conclude that the hole thermionic emission current from metal to semiconductor can be written as

$$J_{pMS} = \frac{qm_p(k_BT)^2}{2\pi^2\hbar^3}\exp(-\psi_{Bp} - \eta_T). \tag{6.105}$$

We reversed the sign, thus accounting for the positive charge of the holes, and used the hole effective mass m_p. (From here to the end of the chapter, the absolute, positive value of the hole effective mass is meant by m_p.) The effective barrier height appearing in the exponential follows from the preceding discussion.

The total hole current density is again the difference of the currents from metal to semiconductor and vice versa:

$$J_{px} = J_{pMS} - J_{pSM} \tag{6.106}$$

(cf. the discussion in Subsection 6.2.4 ahead of Eq. (6.71)). In analogy to (6.70), J_{pSM} results from the formula for J_{pMS} by making the substitution

$$\psi_{Bp} \to \psi_{Bp} - \frac{qV_{cp}}{k_BT} \tag{6.107}$$

in (6.105). The result is

$$\boxed{J_{px} = -\frac{qm_p(k_BT)^2}{2\pi^2\hbar^3}\exp(-\psi_{Bp} - \eta_T)\left[\exp\left(\frac{qV_{cp}}{k_BT}\right) - 1\right].} \tag{6.108}$$

Eq. (6.108) constitutes the boundary condition for the hole continuity equation.

In [30] it has been shown that in typical n-contacts the hole current density at the interface is very much smaller than the electron current (by a factor of about 10^{-4}). Also, qV_{cp} has been found to be small compared to k_BT. Series expansion of (6.108) shows that in this case the condition

$$|J_{px}| \ll \frac{qm_p(k_BT)^2}{2\pi^2\hbar^3}\exp(-\psi_{Bp} - \eta_T) \tag{6.109}$$

holds. Conversely we find that if the hole current is sufficiently small (according to (6.109)), V_{cp} is also small, and by (6.104) the metal Fermi level is nearly equal to the hole quasi-Fermi level. Thus in this case, continuity of the Fermi level at the interface can be used as a boundary condition for the minority carriers [30] instead of (6.108):

$$\phi_p = \phi_M. \tag{6.110}$$

In contrast to (6.108), the condition (6.110) is of Dirichlet type, which is numerically more stable.

However, condition (6.109) may be violated for instance if the contact is located within the minority carrier diffusion length from a *pn*-junction. Under forward bias, minority carriers are injected which do not have the chance to recombine before reaching the contact; a strong minority current at the contact is the result. In this case, a large, unphysical step of the minority quasi-Fermi level at the contact can frequently be observed in simulation results when (6.110) is used as the boundary condition. Under these circumstances, boundary condition (6.108) is to be preferred from a physical point of view.

In the discussion of the contact model, we so far always regarded qV_D as being positive. This sign of qV_D corresponds to a rise of the semiconductor bands when approaching the interface. Such a rise can be caused by the positive space charge of the ionized donors in the depletion region, or by the inversion layer charge of holes at the interface in case of very low doping (see Subsection 6.1.1). The model presented so far is thus valid for contacts on *n*-type semiconductor material, up to the flatband case.

The case of a contact on *p*-type material is shown in Fig. 6.17. The roles of electrons and holes are interchanged once the *p*-doping is sufficiently large. Now the holes are the majority carriers, and the band bending is such that they find a barrier in their way which they might be able to tunnel through. The electrons on the other hand are the minority carriers which are missing a barrier, and which tend to accumulate at the interface.

In summary, if we turn from a contact on *n*-semiconductor to one on *p*-semiconductor, or more precisely from upward to downward band bend-

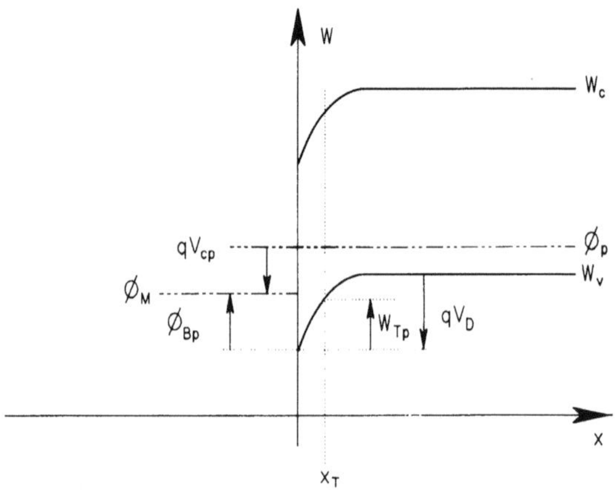

Fig. 6.17 Metal-semiconductor contact on *p*-type semiconductor

ing, we find a symmetric configuration with the roles of electrons and holes interchanged when crossing the flatband case.

Now we extend the previous model to the case of contacts on p-material. For this purpose, we are not going to repeat all the derivations of the previous subsections for the p-case. Instead, we want to make use of the mentioned symmetry in order to rewrite the formulas for the p-type contact by analogy.

First, we introduce some quantities in Fig. 6.17 which will substitute the respective ones in the model equations. The hole barrier height ϕ_{Bp} and the "hole voltage drop" V_{cp} in Fig. 6.17 have already been defined in (6.100) and (6.104). The tunneling energy range W_{Tp} is defined like W_T for the n-contact (cf. Fig. 6.5), but with reversed sign in order to be positive in p-contacts. V_D is not replaced by an analogous quantity; in p-contacts it is negative. The definition of N_D^* also remains the same. Because it is directly related to the band curvature, its sign is also negative in the present case. The quantities corresponding to W_{00} and δ are defined by

$$W_{00p} = \frac{q\hbar}{2} \sqrt{\frac{-N_D^*}{m_p^* \varepsilon_s}} \tag{6.111}$$

and

$$\delta_p = \frac{(k_B T)^2}{-qV_D W_{00p}}, \tag{6.112}$$

respectively.

Thus, the following substitutions have to be made in the model equations for the n-contact in order to arrive at the respective p-contact model:

$$-J_{nx} \to J_{px}, \tag{6.113}$$

$$m_n \to m_p, \tag{6.114}$$

$$\psi_B \to \psi_{Bp}, \tag{6.115}$$

$$V_D \to -V_D, \tag{6.116}$$

$$N_D^* \to -N_D^*, \tag{6.117}$$

$$W_{00} \to W_{00p}, \tag{6.118}$$

$$\delta \to \delta_p, \tag{6.119}$$

$$\eta_T \leftrightarrow \eta_{Tp} = -\eta_T, \tag{6.120}$$

$$V_c \leftrightarrow V_{cp}. \tag{6.121}$$

We shall not write down here the equations that result from these substitutions; instead, a summary of the full model equations is given in the next subsection.

6.2.8 Summary

Since we find the model equations scattered throughout the previous subsections along the derivation, a clearly arranged collection of the equations is presented as a summary in this subsection. Although the summary has been made without any specific implementation in mind, the equations are presented in a sequence as a simulator might evaluate them.

As a first step, we identify the sign of the band bending in order to know if we have tunneling of electrons or holes. Up- or downward bending of the bands is determined by the sign of N_D^*, which is a measure for the band curvature. It is given in Eq. (6.122) in Table 6.3. A positive N_D^* indicates upward band bending; this is the case of electron tunneling which we considered throughout most of Section 6.2. We briefly refer to this case as an *n-contact*. Negative N_D^* indicates downward band bending with hole tunneling. This case will be called *p-contact*. According to the sign of N_D^*, W_{00} is computed with (6.124) using the electron or hole tunneling mass, respectively.

After the determination of the kind of contact, we select the appropriate boundary conditions for the *n*-contact or the *p*-contact from the Tables 6.4 or 6.5, respectively. The discussion proceeds with the *n*-contact case. The *p*-contact case is treated fully analogously.

The Poisson equation boundary conditions are shown first in Table 6.4. The two cases of depletion or neutrality are distinguished. Eq. (6.125) is the boundary condition for the conduction band edge W_c if a *depletion* region is present at the boundary, while (6.126) gives W_c on the boundary in the *neutrality* limit. The function $W_N(\phi_n, \phi_p)$ appearing in this equation is defined in (6.149) in Table 6.8 of auxiliary formulas at the end of the subsection. It arises from the transformation of the original neutrality condition (6.78) into condition (6.81) for W_c.

Table 6.3. *Band curvature parameter; decision between n- and p-contact*

$$N_D^* = \frac{1}{q}\rho_d - \frac{\varepsilon_s}{q^2}\left[\frac{\partial^2 W_c}{\partial y^2} + \frac{\partial^2 W_c}{\partial z^2}\right] \qquad (6.122)$$

$$\rho_d = \begin{cases} q(N_D - N_A), & \text{if } q|N_D - N_A| > |\rho|, \\ \rho(x_T), & \text{otherwise} \end{cases} \qquad (6.123)$$

$$W_{00} = \begin{cases} \dfrac{q\hbar}{2}\sqrt{\dfrac{N_D^*}{m_n^*\varepsilon_s}} & (N_D^* > 0) \\[3ex] \dfrac{q\hbar}{2}\sqrt{\dfrac{|N_D^*|}{m_p^*\varepsilon_s}} & (N_D^* < 0) \end{cases} \qquad (6.124)$$

Table 6.4. *Boundary conditions for the n-contact*

<table>
<tr><td colspan="4" align="center">n-contact ($N_D^* > 0$)</td></tr>
<tr><td colspan="4" align="center">Poisson equation</td></tr>
<tr><td colspan="2" align="center">Depletion
($W_T^{(D)} < W_T^{(N)}$)</td><td colspan="2" align="center">Neutrality
($W_T^{(D)} > W_T^{(N)}$)</td></tr>
<tr><td>$W_c(x_T) = \phi_M + \phi_B - W_T^{(D)}$
with</td><td>(6.125)</td><td>$W_c(x_T) = W_N(\phi_n, \phi_p)$</td><td>(6.126)</td></tr>
<tr><td>$\quad W_T^{(D)} = k_B T \cdot \max(\eta(\psi_B), \eta(\psi_B - \psi_c))$</td><td>(6.127)</td><td>$W_T^{(N)} = \phi_B + \phi_M - W_c(x_T)$</td><td>(6.128)</td></tr>
<tr><td colspan="3">$\eta_T = \dfrac{1}{k_B T} \min(W_T^{(D)}, W_T^{(N)})$</td><td>(6.129)</td></tr>
<tr><td colspan="4" align="center">Continuity equations</td></tr>
<tr><td colspan="3">$J_{nx} = -\dfrac{q m_n (k_B T)^2}{2\pi^2 \hbar^3} [j_T(\psi_B) - j_T(\psi_B - \psi_c)]$</td><td>(6.130)</td></tr>
<tr><td colspan="3">$J_{px} = -\dfrac{q m_p (k_B T)^2}{2\pi^2 \hbar^3} \exp(-\psi_{Bp} - \eta_T)[\exp(\psi_{cp}) - 1]$</td><td>(6.131)</td></tr>
</table>

Table 6.5. *Boundary conditions for the p-contact*

<table>
<tr><td colspan="4" align="center">p-contact ($N_D^* < 0$)</td></tr>
<tr><td colspan="4" align="center">Poisson equation</td></tr>
<tr><td colspan="2" align="center">Depletion
($W_T^{(D)} < W_T^{(N)}$)</td><td colspan="2" align="center">Neutrality
($W_T^{(D)} > W_T^{(N)}$)</td></tr>
<tr><td>$W_c(x_T) = \phi_M + \phi_B + W_T^{(D)}$
with</td><td>(6.132)</td><td>$W_c(x_T) = W_N(\phi_n, \phi_p)$</td><td>(6.133)</td></tr>
<tr><td>$\quad W_T^{(D)} = k_B T \cdot \max(\eta(\psi_{Bp}), \eta(\psi_{Bp} - \psi_{cp}))$</td><td>(6.134)</td><td>$W_T^{(N)} = -\phi_B - \phi_M + W_c(x_T)$</td><td>(6.135)</td></tr>
<tr><td colspan="3">$\eta_T = \dfrac{1}{k_B T} \min(W_T^{(D)}, W_T^{(N)})$</td><td>(6.136)</td></tr>
<tr><td colspan="4" align="center">Continuity equations</td></tr>
<tr><td colspan="3">$J_{nx} = \dfrac{q m_n (k_B T)^2}{2\pi^2 \hbar^3} \exp(-\psi_B - \eta_T)[\exp(\psi_c) - 1]$</td><td>(6.137)</td></tr>
<tr><td colspan="3">$J_{px} = \dfrac{q m_p (k_B T)^2}{2\pi^2 \hbar^3} [j_T(\psi_{Bp}) - j_T(\psi_{Bp} - \psi_{cp})]$</td><td>(6.138)</td></tr>
</table>

Table 6.6. *Formulas for the simultaneous computation of η_T and δ*

Computation of tunneling energy range

$$\eta(\psi_B) = \begin{cases} \sqrt{v_{xm}^*(\psi_B)^2 + \dfrac{3}{\delta}}, & \psi_B < \eta, \ \psi_B < \dfrac{1}{2\delta}, \\[2ex] \sqrt{\dfrac{\psi_B + 3}{\delta} - \dfrac{1}{4\delta^2}}, & \psi_B < \eta, \ \psi_B \geq \dfrac{1}{2\delta}, \\[2ex] \dfrac{1}{2\delta} + \sqrt{\dfrac{3}{\delta}}, & \psi_B \geq \eta \end{cases} \qquad (6.139)$$

$$\frac{1}{\delta} - \frac{W_{00}}{k_B T}\eta(\psi_B) = \frac{\varepsilon_s W_{00}}{2|N_D^*|(k_B T)^2} E_x^2(x_T) \qquad (6.140)$$

$$v_{xm}^*(\psi_B) = -\frac{1}{2}\left[\psi_B - 1 + \sqrt{(\psi_B - 1)^2 + \frac{2}{\delta}} \right] \qquad (6.141)$$

$$E_x(x_T) = \begin{cases} \dfrac{1}{q}\dfrac{\partial W_c}{\partial x}\bigg|_{x=x_T}, & \begin{array}{l} N_D^* > 0 \text{ and } E_x(x_T) < 0 \\ \text{or} \\ N_D^* < 0 \text{ and } E_x(x_T) > 0, \end{array} \\[3ex] 0, & \text{otherwise} \end{cases} \qquad (6.142)$$

According to the discussion at the end of Subsection 6.2.5, the choice between (6.125) and (6.126) in an actual situtation has to be decided such that W_T assumes the smaller value, as indicated in the head of Table 6.4. A practical implementation could first determine W_c from (6.126), then $W_T^{(N)}$ from (6.128). After that, $W_T^{(D)}$ is computed from (6.127) with the help of the formulas in Table 6.6. (Note that a function $\eta(\psi_B)$ has been introduced in (6.139) that summarizes the computation of η_T from ψ_B according to the discussion in Subsections 6.2.5 and 6.2.6.) Then the two W_T's are compared and the boundary condition belonging to the smaller value is selected. For the subsequent evaluation of the tunneling current, η_T is determined by (6.129) and δ by (6.140).

Finally, the boundary conditions for the electron and hole continuity equations, (6.130) resp. (6.131), are displayed. They are written as normal current densities at the boundary. In the case of an n-contact, J_{nx} contains both tunneling and thermionic emission ingredients, while J_{px} describes the thermionic emission of the minority carriers as discussed in Subsection 6.2.7. The formulas for the tunneling current are hidden in the dimensionless function j_T, which is defined in Table 6.7 for degeneration and nondegeneration. Note that the same η_T (and the corresponding δ) has to be used in each of the two evaluations of j_T in (6.143) resp. (6.144). The arguments of the j_T-function in (6.130) and (6.138) are defined in Table 6.8.

Table 6.7. *Formulas for the tunneling and thermionic emission current*

The current function

$\psi < \eta_T$ (degeneration):

$$j_T(\psi) = \frac{1}{2}\sqrt{\frac{\pi}{\delta}}\left\{\exp\left(-\psi + \frac{1}{4\delta}\right)\left[\operatorname{erf}\left(\frac{1}{2\sqrt{\delta}}\right) - \operatorname{erf}\left(\frac{1}{2\sqrt{\delta}} - \sqrt{\delta}\psi\right)\right]\right.$$

$$+ (\psi - 1)[\operatorname{erf}(\sqrt{\delta}\psi) - \operatorname{erf}(\sqrt{\delta}\eta_T)]$$

$$\left. + \frac{1}{\sqrt{\pi\delta}}[\exp(-\delta\psi^2) - \exp(-\delta\eta_T^2)]\right\} + \exp(-\psi) \tag{6.143}$$

$\psi > \eta_T$ (non-degeneration):

$$j_T(\psi) = \frac{1}{2}\sqrt{\frac{\pi}{\delta}}\exp\left(-\psi + \frac{1}{4\delta}\right)\left[\operatorname{erf}\left(\frac{1}{2\sqrt{\delta}}\right) - \operatorname{erf}\left(\frac{1}{2\sqrt{\delta}} - \sqrt{\delta}\eta_T\right)\right]$$

$$+ \exp(-\psi) \tag{6.144}$$

Table 6.8. *Auxiliary formulas*

Auxiliary formulas

Electron and hole barrier height

$$\psi_B = \frac{\phi_B}{k_B T} \tag{6.145}$$

$$\psi_{Bp} = \frac{W_g - \phi_B}{k_B T} \tag{6.146}$$

Electron and hole voltage drop

$$\psi_c = \frac{\phi_n - \phi_M}{k_B T} \tag{6.147}$$

$$\psi_{cp} = \frac{\phi_M - \phi_p}{k_B T} \tag{6.148}$$

Conduction band edge at neutrality

$$W_N(\phi_n, \phi_p) = \frac{1}{2}(\phi_n + \phi_p)$$

$$- k_B T \operatorname{Arsinh}\left(\frac{N_D - N_A}{2n_{ie}}\exp\left[\frac{\phi_p - \phi_n}{2k_B T}\right]\right)$$

$$+ \frac{1}{2}W_g + \frac{k_B T}{2}\ln\left(\frac{N_c \gamma_n}{N_v \gamma_p}\right) \tag{6.149}$$

6.2.9 Discussion

In this subsection, we want to look at some of the features of the contact model. Since the results for the contact current look somewhat complex, we are going to gain an understanding of the model behaviour by consideration of some example calculations.

The relation of the current density to the voltage drop V_c for an Al contact on n-Si with a barrier height of 0.7 V at $T = 300$ K is displayed in Fig. 6.18 for various doping densities. As expected, we note that at lower doping the behaviour of the contact is rectifying. Thermionic emission is dominant in this doping range. At increased doping concentrations, we find a transition to nonlinear resistive curves because of the onset of tunneling current. Further increase of doping density leads to a (nearly) symmetric, ohmic behaviour. Finally, the contact with a doping of 10^{20} cm^{-3} appears as an ideal contact of zero voltage drop on the scale of Fig. 6.18.

The differential specific contact resistance ρ_c is defined by

$$\frac{1}{\rho_c(V_c)} = \frac{\partial J_{nx}}{\partial V_c} \tag{6.150}$$

as the reciprocal slope of the current density [114]. While (6.150) defines the specific resistance as the tangent of the current-voltage-relationship, it is also possible to use the secant of the curve for an alternative definition:

$$\rho_{c,\text{secant}}^{-1}(V_c) = \frac{J_{nx}(V_c)}{V_c}. \tag{6.151}$$

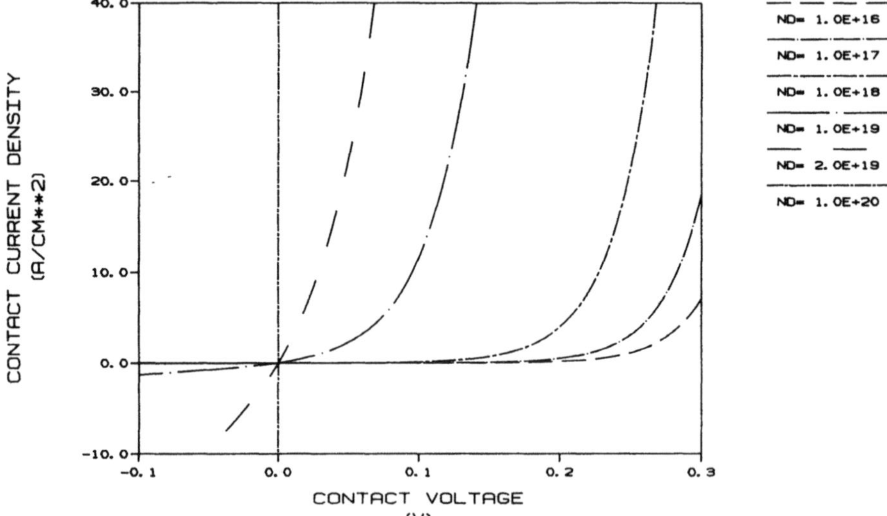

Fig. 6.18 Contact current density versus contact voltage

The specific contact resistance is used in particular for the characterization of ohmic contacts, which have a rather small resistance. The voltage drop V_c is thus commonly small at such contacts. For this reason, ρ_c at $V_c = 0$, the reciprocal slope of the current density at the origin, is of particular importance. If V_c is fairly small, we can linearize the characteristics of Fig. 6.18 at this point, and it suffices to know $\rho_c(0)$ in order to determine the current density through the contact at a given V_c.

Now we want to calculate the specific contact resistance for our contact model. Insertion of (6.130) into (6.150) gives

$$\frac{1}{\rho_c} = -\frac{qm_n(k_B T)^2}{2\pi^2\hbar^3} \cdot \frac{\partial j_T(\psi)}{\partial \psi}\bigg|_{\psi = \psi_B - \psi_c} \cdot \frac{d\psi_c}{dV_c}. \tag{6.152}$$

The latter derivative in (6.152) is simply $q/k_B T$ by the definitions (6.147) and (6.6) for ψ_c and V_c, respectively. With the derivative of the error function

$$\frac{d}{dx} \mathrm{erf}(ax) = \frac{2a}{\sqrt{\pi}} \exp(-a^2 x^2), \tag{6.153}$$

we obtain by differentiation of j_T in (6.143) and (6.144):

$\psi_B - \psi_c < \eta_T$ (degeneration):

$$\frac{1}{\rho_c} = \frac{q^2 m_n k_B T}{2\pi^2\hbar^3} \left\{ \frac{1}{2}\sqrt{\frac{\pi}{\delta}} \exp\left(\psi_c - \psi_B + \frac{1}{4\delta}\right) \right.$$
$$\times \left[\mathrm{erf}\left(\frac{1}{2\sqrt{\delta}}\right) - \mathrm{erf}\left(\frac{1}{2\sqrt{\delta}} - \sqrt{\delta}(\psi_B - \psi_c)\right) \right]$$
$$\left. - \frac{1}{2}\sqrt{\frac{\pi}{\delta}}[\mathrm{erf}(\sqrt{\delta}(\psi_B - \psi_c)) - \mathrm{erf}(\sqrt{\delta}\eta_T)] + \exp(\psi_c - \psi_B) \right\} \tag{6.154}$$

$\psi_B - \psi_c > \eta_T$ (non-degeneration):

$$\frac{1}{\rho_c} = \frac{q^2 m_n k_B T}{2\pi^2\hbar^3} \left\{ \frac{1}{2}\sqrt{\frac{\pi}{\delta}} \exp\left(\psi_c - \psi_B + \frac{1}{4\delta}\right) \right.$$
$$\left. \times \left[\mathrm{erf}\left(\frac{1}{2\sqrt{\delta}}\right) - \mathrm{erf}\left(\frac{1}{2\sqrt{\delta}} - \sqrt{\delta}\eta_T\right) \right] + \exp(\psi_c - \psi_B) \right\} \tag{6.155}$$

as the determinative relations for the specific contact resistance.

Fig. 6.19 shows the specific contact resistance according to (6.154)–(6.155) at $V_c = 0$ for selected contact materials as a function of doping concentration together with respective experimental data. Experimental data and barrier heights of 0.58 eV, 0.7 eV, and 0.85 eV for Nb-Si, Al-Si, and Pt-Si contacts have been taken from [117], [114], and [42], respectively. Although no parameter adjustment has been made, the agreement is quite

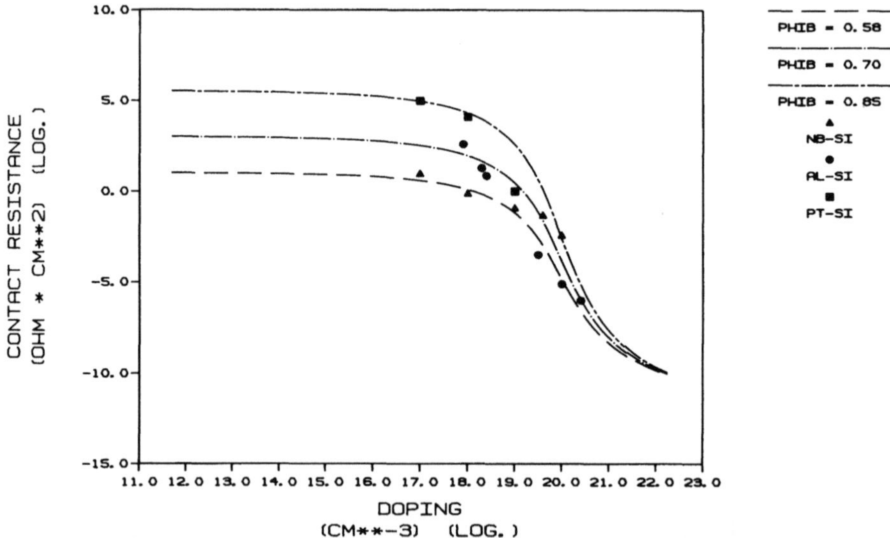

Fig. 6.19 Specific contact resistance versus doping concentration ($T = 300$ K)

reasonable. At low doping, we find an approximately constant contact resistance. This is the thermionic emission regime, where δ tends to large values, and consequently the first term in (6.155) becomes negligible. As the doping increases above 10^{18} cm^{-3}, we note a drastic decrease of the resistance due to the onset of tunneling.

As an alternative, the contact current can be equivalently expressed in terms of a *recombination velocity* v_n by writing

$$J_{nx} = qv_n(n(x_T) - n_0).\tag{6.156}$$

The term "recombination velocity" has been introduced into the theory of metal-semiconductor contacts in a frequently cited paper by Crowell and Sze in 1966 [79].

The formulation (6.156) is obtained from (6.71) by defining

$$v_n = \frac{J_{SM}}{-qn(x_T)}\tag{6.157}$$

and

$$n_0 = n(x_T)\frac{J_{MS}}{J_{SM}}.\tag{6.158}$$

These expressions become particularly simple in the case of *non-degeneration*. For this reason, the recombination velocity formulation is especially useful for Schottky contacts and will be discussed in more detail in Section 6.3.

In order to determine the parameters of (6.156), the semiconductor electron concentration at the boundary $n(x_T)$ is required. With (2.14), (6.6), (6.9), and (6.8) we find

$$n(x_T) = N_c \exp(-\psi_B + \eta_T + \psi_c). \tag{6.159}$$

Insertion into (6.157) gives with (6.65) the recombination velocity

$$v_n = \sqrt{\frac{k_B T}{2\pi m_n}} \exp(-\eta_T) \times \left\{ 1 + \frac{1}{2}\sqrt{\frac{\pi}{\delta}} \exp\left(\frac{1}{4\delta}\right) \right.$$
$$\left. \times \left[\operatorname{erf}\left(\frac{1}{2\sqrt{\delta}}\right) - \operatorname{erf}\left(\frac{1}{2\sqrt{\delta}} - \sqrt{\delta}\eta_T\right) \right] \right\}. \tag{6.160}$$

From (6.65) with (6.70), we obtain $J_{MS}/J_{SM} = \exp(-\psi_c)$, and (6.158) becomes with (6.159)

$$n_0 = N_c \exp(-\psi_B + \eta_T). \tag{6.161}$$

Note that through η_T and δ, the recombination velocity and n_0 still depend on the contact voltage drop. The second term in the curly brackets in (6.160) accounts for tunneling. For vanishing tunneling at lower doping concentrations, η_T goes to zero, and v_n and n_0 turn into the respective values of Schottky contact theory (see Section 6.3).

Throughout this chapter, we concentrated only on thermionic emission and tunneling, disregarding the other transport effects mentioned in Section 6.1. However, a final remark on surface recombination is in order.

If the interface between metal and semiconductor has a large number of interface states, recombination through these states may become important [118]. This effect can be considered by superposition of the thermionic emission and tunneling currents flowing across the contact and the recombination current passing through the interface traps [118]. The boundary condition for the hole continuity equation at the n-contact, for instance, then becomes the sum of (6.131) and (5.27). Other combinations are analogous.

6.3 Schottky Contact

In this section, we consider the low doping limit of the general contact model derived in the previous section. As discussed qualitatively in Section 6.1, the bands bend only slowly if doping is low. The effect is that the barrier is thick in this case and the tunneling probability goes to zero. Hence, thermionic emission is the dominating current transport mechanism, and the contact has a rectifying behaviour. The device which employs the rectification effect of a contact like this is the "Schottky diode". Another application of a Schottky contact is the controlling gate in a metal-semiconductor field-effect transistor (MESFET).

In the following, we explain the low doping limit only for the case of an n-contact. The corresponding equations for a p-contact are fully analogous. In the model of Section 6.2, the parameter describing the band curvature in the direction normal to the interface is N_D^*, defined in (6.122). Small band curvature means that N_D^* is small. Correspondingly, by (6.124), also W_{00} goes to zero. From (6.140) follows that δ becomes large, and from (6.139) that η_T as well as W_T become small. In the asymptotic limit we get

$$W_T = 0 \qquad \text{(no tunneling)}. \tag{6.162}$$

Recalling that W_T can be viewed as an apparent barrier lowering due to tunneling, the obvious interpretation of (6.162) is that the tunnel effect becomes negligible in this limit. The carriers have to overcome the full barrier height when crossing the interface. Also, the tunneling length x_T goes to zero, and the position of the boundary is at the peak of the barrier.

We obtain the boundary condition for the *Poisson equation* in this limit by inserting (6.162) into the Poisson boundary condition (6.125). The result is

$$\boxed{W_c(x_T) = \phi_M + \phi_B.} \tag{6.163}$$

It specifies the conduction band edge as a function of the barrier height and the metal Fermi level, i.e. the voltage applied to the contact (cf. Section 6.2.5). The neutrality boundary condition of (6.126) is not applicable, since the edge of the depletion region in a Schottky contact is relatively far away from the interface in the bulk of the semiconductor.

We derived the Schottky contact boundary condition for the Poisson equation by specifying the conduction band edge at the boundary. In many practical implementations, a different but equivalent formulation is used which instead specifies the intrinsic Fermi energy at the boundary. Inserting (6.163) into (3.24), we obtain

$$W_i(x_T) = \phi_M + W_S, \tag{6.164}$$

where W_S is given by

$$W_S = \phi_B - \frac{W_g}{2} + \frac{k_B T}{2} \ln\left(\frac{N_v}{N_c}\right). \tag{6.165}$$

Dividing by q, these equations can be written likewise for potentials (see e.g. [119]). Hence, the intrinsic potential at the boundary equals the applied voltage, plus the contact specific "surface potential" W_S/q [119]. Eq. (6.163) reflects the physical situation somewhat clearer than (6.164), where the effect of the Schottky barrier is obscured by combining ϕ_B with the terms from the definition of the intrinsic Fermi energy (3.24) into W_S.

Next we consider the boundary condition for the electron continuity equation. Starting from the general condition (6.130), we note that in the definition of j_T in (6.143) and (6.144), all terms except the last one in each case have a common factor $1/\delta$ in front. We remember that these terms describe the tunneling current, while the last term describes the thermionic emission current. Consequently, the tunneling terms vanish in the present case, since $1/\delta$ goes to zero. Only the thermionic emission term remains. Inserting (6.144) into (6.130) for this limit, we obtain

$$J_{nx} = \frac{qm_n(k_B T)^2}{2\pi^2\hbar^3} \exp(-\psi_B)[\exp(\psi_c) - 1]$$

(6.166)

as the boundary condition for the *electron continuity equation*. Clearly, the boundary condition specifies the normal component of the electron current density as a function of the barrier height and the difference of the Fermi levels in the metal and the semiconductor. Since no tunneling occurs, the boundary condition for the continuity equation in this case is identical to the well-known formula for the thermionic emission current [30]. Eq. (6.166) is the boundary condition for the Schottky contact commonly used in most device simulators. The barrier height in (6.166) depends in general on the electric field at the contact because of image force lowering, which leads to increased current densities particularly at strong reverse bias.

Many authors like to combine the prefactor in (6.166)—excluding the temperature—into a single parameter

$$A_n^* = \frac{qm_n k_B^2}{2\pi^2\hbar^3},$$

(6.167)

which is known as the effective "Richardson constant" [78, 120]. Accepted values for A_n^* are 110 and 8 $AK^{-2}cm^{-2}$ for Si and GaAs, respectively [120]; the corresponding values of A_p^* for holes are 32 (p-Si) and 74 $AK^{-2}cm^{-2}$ (p-GaAs) [120].

Another reformulation of (6.166) is very common in the literature. At first glance, it looks like something completely different, but we will see shortly that the formula is fully equivalent. In this formulation, the contact current density is specified in terms of the recombination velocity v_n as introduced by Crowell and Sze [79] (see also Section 6.2.9). The boundary condition reads

$$J_{nx} = qv_n(n(x_T) - n_0),$$

(6.168)

where $n(x_T)$ is the electron concentration at the boundary (we keep the notation x_T for the boundary coordinate, although tunneling is no longer present here). It is given by

$$n(x_T) = N_c \exp\left(-\frac{W_c(x_T) - \phi_n(x_T)}{k_B T}\right),$$

(6.169)

according to (2.14). The quantity n_0 is an abbreviation defined by

$$n_0 = N_c \exp\left(-\frac{\phi_B}{k_B T}\right). \tag{6.170}$$

It has the dimension of a concentration, and is referred to as the "quasi-equilibrium concentration", since it is a constant equal to $n(x_T)$ at zero current. Finally, the recombination velocity is given by

$$v_n = \sqrt{\frac{k_B T}{2\pi m_n}} = \frac{q m_n (k_B T)^2}{2\pi^2 \hbar^3} \frac{1}{q N_c}. \tag{6.171}$$

Occasionally, it is also referred to as "thermionic emission velocity", which better reflects the physical nature of the process. It results from (6.160) in the limit $\delta \to \infty$ and with $\eta_T = 0$. The latter equality in (6.171) becomes obvious when we recall the definition (2.15) for the effective density of states. In the formulation of (6.168), the boundary condition is interpreted as the net flux of two concentrations n and n_0, streaming in opposite directions with the same absolute velocity v_n.

Our original formulation (6.166) is recovered if we insert (6.169), (6.170), and (6.171) into (6.168), which gives

$$J_{nx} = \frac{q m_n (k_B T)^2}{2\pi^2 \hbar^3} \frac{1}{N_c} \left[N_c \exp\left(-\frac{W_c(x_T) - \phi_n(x_T)}{k_B T}\right) \right.$$
$$\left. - N_c \exp\left(-\frac{\phi_B}{k_B T}\right) \right]. \tag{6.172}$$

Taking out a factor of $\exp(-\psi_B)$ (we used again the definition (6.145) for the normalized barrier height ψ_B), we get for the first exponent in (6.172)

$$-\frac{W_c(x_T) - \phi_n(x_T) - \phi_B}{k_B T} = -\frac{\phi_M - \phi_n(x_T)}{k_B T}, \tag{6.173}$$

where we substituted ϕ_B on the right-hand side by its definition (6.1). Rewriting the Fermi level difference with the help of (6.147), we arrive again at the original formulation (6.166). Thus we have shown that the general contact model of Section 6.2 contains the Crowell-Sze model as a special case. On the other hand, we see that the Crowell-Sze model too is based on the assumption of the discontinous interface distribution (6.30).

Finally, the boundary condition for the *hole continuity equation* for the Schottky contact is considered. With $\eta_T = 0$, the general condition (6.131) becomes

$$\boxed{J_{px} = -\frac{q m_p (k_B T)^2}{2\pi^2 \hbar^3} \exp(-\psi_{Bp})[\exp(\psi_{cp}) - 1].} \tag{6.174}$$

This equation describes the thermionic emission of holes across the metal-semiconductor interface. Eq. (6.174) likewise can be rewritten into the formulations employing the Richardson constant or the recombination velocity.
If the injection of minority carriers at the contact is negligible, the boundary condition can be simplified to the small current limit $\psi_{cp} = 0$ of (6.174)

$$\boxed{\phi_p = \phi_M.} \tag{6.175}$$

The bulk of the papers that established our present understanding of metal-semiconductor contacts appeared in the years around 1965–1975. However, since device simulation at that time was in its early beginnings, the theoretical considerations then concentrated on the current-voltage relation of the total contact, including the interface as well as the depletion region. The early *emission theory* of Bethe described the carrier transport at the contact as thermionic emission across the barrier and the depletion region into the neutral zone; scattering in the depletion region was neglected. Another frequently discussed concept was the *diffusion theory*. The diffusion theory tried to explain the transport process by drift-diffusion in the depletion region, assuming continuity of the Fermi level at the metal-semiconductor interface as a boundary condition. With the *thermionic emission-diffusion theory* (also described basically in an earlier paper by Schultz [121]), Crowell and Sze introduced a combination of the two theories, which considered thermionic emission at the barrier and drift-diffusion in the depletion region as two effects in series [79]. In this theory, (6.168) was used as a boundary condition for the analytical solution of the drift-diffusion equations in the depletion region. The result was a formula for the current-voltage characteristics of the Schottky diode. It was a unification of the emission and the diffusion theory, in that it transforms into these theories for the limits of high resp. low electron mobility.
In device simulation, we have a slightly different situation. Now the differential equation solver takes care of the carrier transport in the semiconductor, including the depletion region near the contact. Only in the immediate vicinity of the interface, where the transport equations of the semiconductor bulk become invalid, special action has to be taken. We have done this by placing the boundary of the simulation domain such that the transport equations are valid in the whole simulation domain, while the effects near the interface are incorporated in the boundary conditions. Thus, in device simulation the question for emission or diffusion theory simply does not arise, since all we need is a boundary condition describing the interface effects, and the rest of the device is taken care of by the numerical solver of the transport equations.
In addition to the introduction of the recombination velocity, Crowell and Sze [79] discussed scattering of the electrons during interface transfer by

optical phonons and presented some corresponding numerical results. Also quantum-mechanical reflection and tunneling was briefly considered by introduction of a corresponding transition probability factor for the recombination velocity. Practical formulas, which would be suitable for device simulation, however were not given. In a later paper [122], Crowell and Beguwala pointed out that the Fermi level is discontinous under the boundary condition (6.168), which is clear from the discussion in Subsection 6.2.2.

Following the initial basic publications on the modeling of metal-semiconductor contacts, several authors advanced the theory by considering the distribution in the vicinity of the interface in more detail.

Baccarani obtained an analytical expression for the distribution function in the transition region by an approximate solution of the Boltzmann equation [113]. He found an exponential decay of the interface perturbation within the mean free path, which was caused by scattering events.

Berz [123] showed that in absence of scatterings and in the strong forward regime a transition from a hemi-Maxwellian distribution at the interface to an almost full Maxwellian distribution in the "bulk" occurs over a distance where the band edge changes by a value of about $k_B T$. The distance of this interface layer can be much shorter than the mean free path if the the electric field is sufficiently strong. This behaviour is contained as a special case in Baccarani's result.

Other authors accounted for the behaviour of the distribution function by considering distributions different from the hemi-Maxwellians (6.27) and (6.31) when deriving the boundary condition. In particular, the 2nd alternative of the assumptions listed in Subsection 6.2.2, i.e. the volume distribution, has been adopted.

Adams and Tang [124] derived a boundary condition for the electron continuity equation at a Schottky contact by taking a shifted Maxwellian distribution

$$f_2(\mathbf{k}_2) = \exp\left(-\frac{1}{k_B T} \left[\frac{\hbar^2 (\mathbf{k}_2 - \mathbf{k}_d)^2}{2m_n} + W_{cS} - \phi_n \right] \right) \tag{6.176}$$

in the semiconductor instead of the two halves of an equilibrium distribution as in (6.30). The shift $\mathbf{k}_d$ in $\mathbf{k}$-space describes a current density with the drift velocity $\mathbf{v}_d = \hbar \mathbf{k}_d / m_n$. Adams and Tang then *assumed*—without further justification—a recombination velocity defined by

$$v_n = \frac{\int_0^\infty dk_x \, v_x f_2}{\int_0^\infty dk_x \, f_2}, \tag{6.177}$$

which can be interpreted as the mean velocity of electrons travelling in the positive x-direction. (Comparing (6.177) with (6.157), we note a factor of two between the definitions.) Inserting (6.176) into (6.177), they obtained (with the sign of v_{dx} adapted to the reverse arrangement of metal and semi-conductor in [124])

$$v_n = 2v_{n0} \frac{\exp\left(-\dfrac{m_n v_{dx}^2}{2k_B T}\right)}{1 - \mathrm{erf}\left(v_{dx}\sqrt{\dfrac{m_n}{2k_B T}}\right)} - v_{dx}. \tag{6.178}$$

In this equation, v_{dx} is the x-component of the drift velocity $\mathbf{v}_d$, which is determined during the simulation from $\mathbf{v}_d = -\mathbf{J}_n/(qn)$. The velocity v_{n0} is given by

$$v_{n0} = \sqrt{\frac{k_B T}{2\pi m_n}}, \tag{6.179}$$

which is identical to the constant recombination velocity (6.171) of the Crowell-Sze theory. The result (6.178), inserted in (6.168), has been used in device simulations as a Schottky contact boundary condition [125].

In a subsequent paper, Nylander et al. [119] modified this boundary condition by introducing a factor η in front of m_n in (6.176), (6.178) and (6.179), which was fitted to $\eta = 4$ such that the recombination velocity in equilibrium ($\mathbf{v}_d = 0$) becomes equal to v_{n0}. This modification removed the factor 2 in front of v_{n0} in (6.178), and thus restored the original Crowell-Sze recombination velocity in equilibrium. The authors used the current dependent recombination velocity only for forward bias ($v_{dx} < 0$); for reverse bias $v_n = v_{n0}$ has been taken.

It is interesting to see what the inflow moments method yields for the shifted-Maxwellian assumption (6.176). Instead of (6.30) with (6.31), we insert (6.176) into the boundary condition (6.26). In the metal, we assume as before the equilibrium distribution (6.29). Evaluating (6.26) for thermionic emission, we find the contact current density

$$J_{nx} = qv_n n(x_T) - qv_{n0} n_0, \tag{6.180}$$

where v_n is now given by

$$v_n = v_{n0} \exp\left(-\frac{m_n v_{dx}^2}{2k_B T}\right) - \frac{1}{2} v_{dx}\, \mathrm{erfc}\left(v_{dx}\sqrt{\frac{m_n}{2k_B T}}\right), \tag{6.181}$$

and n_0 is defined in (6.170). This recombination velocity has been derived before by Darling [126]. It should be compared to (6.178). The contact model (6.180)–(6.181) has the advantages that the recombination velocity equals v_{n0} for zero bias and approaches $|v_{dx}|$ for large forward bias (as in the formulation of Nylander et al. [119]), while it interpolates smoothly between forward and reverse bias. Moreover, the model has been rigorously derived by calculating the current density instead of simply assuming

(6.177). Also, it is physically more plausible in (6.180) that the current injected from the metal is independent of v_{dx}.

Hot-electron effects at metal-semiconductor contacts can be accounted for by including the energy balance equation into the analysis. In this case, an additional boundary condition for the energy density—or likewise the electron temperature—is necessary. This problem has been treated in analytical solutions by Stratton [127] and Darling [126], and for device simulation by Hjelmgren [128].

Stratton disregarded thermionic emission at the contact, but used a very simple boundary condition by setting the electron temperature to the lattice temperature. While Darling arrived at a thermionic emission current formula which accounts for a non-equilibrium electron temperature, he did not give a boundary condition for the energy balance equation. Hjelmgren, noting that it is inappropriate to force the electrons to equilibrium *before* they cross the interface, obtained thermionic emission boundary conditions for both the particle and the energy balance equation.

6.4 Ohmic Contact

If the doping of the semiconductor is high, band bending is large, and the barrier at the metal-semiconductor interface is very thin. Tunneling is the dominant transport mechanism, which leads to high current densities at low voltage drops. Consequently, the resistance of the contact is very low, and the current-voltage characteristic can be approximately taken as linear. This kind of contact is called *ohmic contact*. It is used in semiconductor devices for the application of external voltages and the supply of current from external sources. This section introduces the models for the ohmic contact outgoing from the high doping limit of the general contact model of Section 6.2. We consider electrons as the majority carriers; the models for contacts on p-type semiconductor can be obtained by analogy.

We begin with the boundary condition for the Poisson equation. At high doping concentrations, the tunneling length x_T comprises the total depletion region, and the position of the boundary is placed at the depletion layer edge (cf. the discussion in Subsection 6.2.5). The conduction band minimum at the depletion layer edge is determined by the charge neutrality condition. Thus, we get as the boundary condition for the *Poisson equation*

$$\boxed{n - p - N_D + N_A = 0.} \tag{6.182}$$

This is the commonly used boundary condition at ohmic contacts. It is obtained here as the high doping limit of the general model. Recall that we can equivalently use (6.126) in Table 6.4 in order to have a boundary condition directly for the conduction band edge.

Considering now the boundary condition for the *electron continuity equation*, we note that the voltage drop $\phi_n - \phi_M$ at the contact is usually small, since the tunneling effect is able to support high current densities with small Fermi level differences across the contact. Hence, we may linearize the current density expression (6.130) at $\psi_c = 0$. The current density then becomes

$$J_{nx} = \frac{1}{q\rho_c}(\phi_n - \phi_M) \qquad \text{(resistive contact)}, \tag{6.183}$$

where the specific contact resistance ρ_c is defined in (6.150) with V_c taken to be zero. Approximate expressions for ρ_c are derived below. It will be shown that ρ_c tends to extremely small values at very high doping. In the limit $\rho_c \to 0$, (6.183) reduces to the boundary condition for the *ideal ohmic contact*:

$$\phi_n = \phi_M \qquad \text{(ideal contact)}. \tag{6.184}$$

Only if (6.184) is satisfied, the current density stays finite when ρ_c goes to zero. Eq. (6.184) is the condition which is used (in a more or less rewritten form) for ohmic contacts in almost every device simulator.

Finally, we turn to the minority carriers, i.e. the boundary condition for the *hole continuity equation*. In principle, the thermionic emission current density (6.131) has to be used. In the highly doped ohmic contacts, however, the minority carrier concentration is particularly small. Thus, the discussion in Subsection 6.2.7 concerning the smallness of the minority current density is all the more valid for ohmic contacts, and according to (6.110) we may take the continuous Fermi level approximation

$$\phi_p = \phi_M \tag{6.185}$$

as the boundary condition for the hole continuity equation. Eq. (6.185) completes the ohmic contact boundary conditions that are commonly used in current implementations of device simulation programs.

The boundary condition for the Poisson equation can be rewritten in case of the ideal contact by insertion of (6.184) and (6.185) into (6.149); the result is

$$W_c = \phi_M - k_B T \, \text{Arsinh}\left(\frac{N_D - N_A}{2n_{ie}}\right) + \frac{1}{2} W_g$$

$$+ \frac{k_B T}{2} \ln\left(\frac{N_c \gamma_n}{N_v \gamma_p}\right). \tag{6.186}$$

If we assume—in a modelling approach—an ideal contact on a non-degenerate semiconductor, the boundary condition can be particularly simply expressed in terms of the intrinsic Fermi energy on the boundary; with the help of (3.24), we obtain from (6.186)

$$W_i = \phi_M - k_B T \, \text{Arsinh}\left(\frac{N_D - N_A}{2n_{ie}}\right). \tag{6.187}$$

In the following, we derive approximate expressions for the specific contact resistance at $\psi_c = 0$. For this purpose, we simplify (6.154) for the case of very high doping. Eq. (6.155) does not need to be considered, since the doping in ohmic contacts is always so high that the electrons in the semiconductor are in degeneration. As explained in detail in Subsection 6.2.6, δ goes to zero as the doping concentration becomes high. Thus we obtain a high-doping approximation of (6.154) if we look for the asymptotic limit of the formula for $\delta \to 0$.
Both of the tunneling terms in (6.154) contain differences of error functions. These differences tend to zero for small δ, since the individual terms of the differences approach one another. With the asymptotic expansion of the error function

$$\text{erf}(x) \approx 1 - \frac{1}{\sqrt{\pi}} \frac{1}{x} \exp(-x^2) \qquad (x \gg 1) \tag{6.188}$$

for large arguments, we get the following approximation of the first difference expression for small δ:

$$\text{erf}\left(\frac{1}{2\sqrt{\delta}}\right) - \text{erf}\left(\frac{1}{2\sqrt{\delta}} - \sqrt{\delta}\psi_B\right)$$

$$\approx 2\sqrt{\frac{\delta}{\pi}} [\exp(-\delta\psi_B^2) - \exp(-\psi_B)] \exp\left(\psi_B - \frac{1}{4\delta}\right). \tag{6.189}$$

Taylor expansion of the second difference in (6.154) around $\sqrt{\delta}\psi_B$ yields

$$\text{erf}(\sqrt{\delta}\psi_B) - \text{erf}(\sqrt{\delta}\eta_T) \approx 2\sqrt{\frac{\delta}{\pi}}(\psi_B - \eta_T)\exp(-\delta\psi_B^2). \tag{6.190}$$

Inserting these expressions into (6.154) and taking the reciprocal, we obtain

$$\rho_c = \frac{2\pi^2\hbar^3}{q^2 m_n (W_T - \phi_B + k_B T)} \exp(\delta\psi_B^2), \tag{6.191}$$

where the normalized quantities η_T and ψ_B have been replaced with W_T and ϕ_B using (6.129) and (6.145), respectively.
If we now investigate the doping dependence of ρ_c, we first note that at neutral, highly doped contacts, the electron density n is nearly equal to the doping density N_D, where n is related to the quasi-Fermi level by (see (2.22))

$$n = N_c F_{1/2}\left(\frac{\phi_n - W_c}{k_B T}\right). \tag{6.192}$$

Furthermore, we observe in (6.191) that

$$W_T - \phi_B = \phi_n - W_c, \tag{6.193}$$

which follows from (6.128) and $\phi_n \approx \phi_M$. With (6.192), W_T in (6.193) can be expressed by n and hence by N_D. Solving (6.192) for $\phi_n - W_c$ and inserting into (6.193), we obtain

$$W_T - \phi_B = k_B T F_{1/2}^{-1}\left(\frac{N_D}{N_c}\right). \tag{6.194}$$

With the approximation of the Fermi integral in degeneration [19]

$$F_{1/2}(x) \approx \frac{4}{3\sqrt{\pi}} x^{3/2} \qquad (x \gg 1), \tag{6.195}$$

we can rewrite (6.194) for W_T

$$W_T = \phi_B + \phi_n - W_c = \phi_B + k_B T \left(\frac{3\sqrt{\pi}}{4}\frac{N_D}{N_c}\right)^{2/3}. \tag{6.196}$$

This relation gives us the basic doping dependence of the denominator in (6.191).

The dominating doping dependence of the exponent in (6.191) becomes

$$\delta\psi_B^2 = \frac{\phi_B^2}{W_T W_{00}} \sim N_D^{-\alpha}, \tag{6.197}$$

where $E_x = 0$ in (6.140) for the neutral contact has been regarded. With $W_{00} \sim N_D^{1/2}$ from (6.124), the value of α is either $\frac{1}{2}$ or $\frac{7}{6}$, according to which of the two terms in (6.196) dominates.

Sze [120] derived a formula for the specific contact resistance by a rough estimation of the tunneling probability in the limit of weak tunneling using a square potential. He obtained a tendency of the doping dependence in surprising agreement with (6.197) for $\alpha = \frac{1}{2}$, which was shown to agree with experimental data in the respective doping range.

From the above discussion we find that ρ_c approaches with a factor of $\exp(N_D^{-\alpha})$ the slowly varying asymptotic

$$\rho_{c,\min} = \frac{2\pi^2 \hbar^3}{q^2 m_n (W_T - \phi_B + k_B T)}. \tag{6.198}$$

With (6.196), (6.198) can be expressed as a function of N_D:

$$\rho_{c,\min} = \frac{2\pi^2 \hbar^3}{q^2 m_n k_B T\left(1 + \left(\frac{3\sqrt{\pi}}{4}\frac{N_D}{N_c}\right)^{2/3}\right)}. \tag{6.199}$$

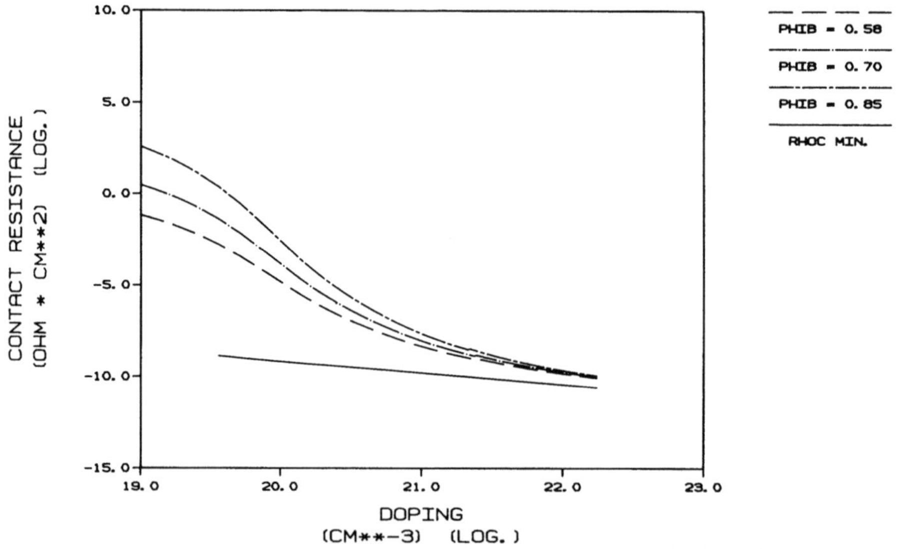

Fig. 6.20 Specific contact resistance at very high doping levels

Note that $\rho_{c,\min}$ depends only on semiconductor material parameters and the electron concentration at the boundary. In particular, it is independent of the barrier height ϕ_B, and hence independent of the metal species of the contact. The value of $\rho_{c,\min}$ is the minimum specific contact resistance that can be achieved with a given semiconductor.

The asymptotic behaviour of ρ_c is depicted in Fig. 6.20. The doping dependence of the specific contact resistance for the barrier heights of Fig. 6.19 is shown together with the asymptotic (6.199). This approach of ρ_c towards a barrier height-independent minimum value, which changes only slighly with increasing doping concentration, has been confirmed by numerical solutions of a more sophisticated contact model [129]. The explanation of this behaviour is that the barrier becomes so thin at some sufficiently high doping level, that the transmission probability approaches unity. The barrier then is fully transparent to the electrons, and the barrier height is of no concern anymore. A further increase of the doping concentration then is nearly ineffective with respect to a decrease of the contact resistance. Indeed, Kupka and Anderson [130] arrived at essentially the same expression for $\rho_{c,\min}$ by assuming $D = 1$ in the current integral right from the beginning.

Early investigations of tunneling dominated metal-semiconductor have been undertaken by Padovani and Stratton [104], based on former work of Stratton [110]. They assumed a parabolic barrier and the WKB approximation, and approximated the current integral by several expansions for the different cases of field emission, thermionic field emission, degeneration, non-degeneration, and forward and reverse current. This model was very complicated and moreover not fully consistent, since it failed to give zero current in equilibrium for some cases.

The treatment of Crowell and Rideout [83] was based on similar assumptions. An analytical expression for the tunneling probability was obtained (see (6.13)). However, the model was limited to the non-degenerate case, and only numerical results for the currents were given. It was thus not suitable to be used as a boundary condition in device simulations.

Pioneering work on the understanding of ohmic contacts in semiconductor devices has been undertaken by Yu [114]. He derived approximate expressions for the specific contact resistance based on the work of Padovani. He also pointed out the importance of the doping level for the dominance of thermionic or tunnel emission.

Shenai [107] reported a comprehensive study of transport at ohmic contacts, which took into account a lot of effects that earlier studies had to neglect or approximate. These included metal-induced gap states, mixing of conduction and valence states, space charge of tunneling electrons, and a more realistic shape of the potential barrier. Of course, only a numerical solution of this problem was possible.

In the context of device simulation, the ideal ohmic contact model as described above has been used from the early beginnings [97], [98], [94], [95] until today [115, 48, 131]. Its simplicity and good numerical properties [132] make it a choice whenever the properties of the ohmic contact are not essential for overall device behaviour.

Modern simulators, which account for hot-electron effects by inclusion of the energy balance equation, usually assume as the additional boundary condition for this equation that the electron temperature at the contact equals the lattice temperature [133]. Recently, device simulators which determine the hot-electron energy distribution using a spherical harmonics expansion of the Boltzmann equation are under development [134, 135]. As a boundary condition, the assumption of an equilibrium distribution at the ideal contact usually has been employed. An improved model that better accounts for the collection of hot electrons at the contact has been published in [136].

7 Semiconductor Heterojunction

Semiconductor heterojunctions are interfaces where two different semi-conducting materials are in contact. This definition also includes transition surfaces where the composition of an alloy semiconductor changes abruptly, as e.g. a $Ga_{0.6}Al_{0.4}As$–$Ga_{0.9}Al_{0.1}As$ interface. If the heterojunction (sometimes also termed heterointerface) has the same type of majority carriers (n-n or p-p) on both sides of the interface, it is called "isotype". If the majority carriers are electrons on one side and holes on the other side, the heterojunction is called "anisotype" [87].

Semiconductor heterojunctions have been a research topic since many years [28], [87], [137]. Recently, technical applications of heterojunctions in novel device structures have become increasingly important. In particular, the advent of molecular beam epitaxy (MBE) technology has made possible a precise control of the composition and variation of the different semiconductor layers. This technique allows to define the energy band model of a device, i.e. the position dependence of the magnitude of the bandgap and the line-up of the bands in an unprecedented way. Also the local rate of change of composition can be controlled precisely, leading to either gradual or abrupt heterojunctions. On this grounds, a new philosophy of semiconductor device design, *bandgap-engineering*, has developed. A comprehensive review of the physical mechanisms, models, and measurement techniques of energy band discontinuities, as well as the physics of heterojunction devices has been given in [137].

In many heterojunction devices, the usage of band discontinuities is functionally employed. For example, steep barriers can be constructed that represent an obstacle to most (i.e. low energetic) electrons. Only hot carriers with high kinetic energies are able to cross the barrier and enter the high energy side. Because carriers are transferred to a locally adjacent semiconductor region, this effect is called *Real-Space Transfer* (RST) [138]. If the adjacent region has a lower mobility, the RST can give rise to a negative differential resistance [139]. The name RST has evolved because of the similarity of the effect with momentum-space transfer, which is also a possible reason for a negative differential resistance when the electrons transfer in k-space to a valley with higher effective mass.

Another utilization of band discontinuities is the confinement of electrons in a narrow well. The well can be undoped and thus can have a very small impurity scattering rate; as a consequence, the electron mobility in the well can be very high. The electron concentration in the well can be controlled by a nearby gate electrode; this is the principle of the *High Electron Mobility Transistor* (HEMT) [63].

Abrupt heterojunctions are also used for the injection of electrons with a high kinetic energy into a semiconductor region (electron launcher) [140, 141]. Here, electrons are accelerated on the high-energy side of the heterojunction towards a steep chasm (see Fig. 7.1), where they finally fall down over the edge and enter the low region with a very high velocity.

As a last example, the utilization of band edge discontinuities for very thin barriers shall be mentioned. The barriers are so thin that the tunneling probability is considerably large. If several such barriers are positioned in series, a resonant structure emerges that allows electrons with certain energies to transfer much more easily than those with different energies. This is the *Resonant Tunneling* effect [142].

For simulations of actual heterojunction devices, numerical values of the band edge discontinuities are required. Reliable experimental data for every possible combination of materials are presently not available. Margaritondo collected a large database of numerous experimental data, which can be found in [47]. Also limitations and accuracy issues are discussed in that reference.

Instead of repeating Margaritondo's table, I prefer to present at this point a practical method (also discussed in [47]) to estimate the band edge discontinuities of heterojunctions from Tersoff's theory of band lineups (see Chapter 3). Recall that the conduction band step was given by (3.22) as

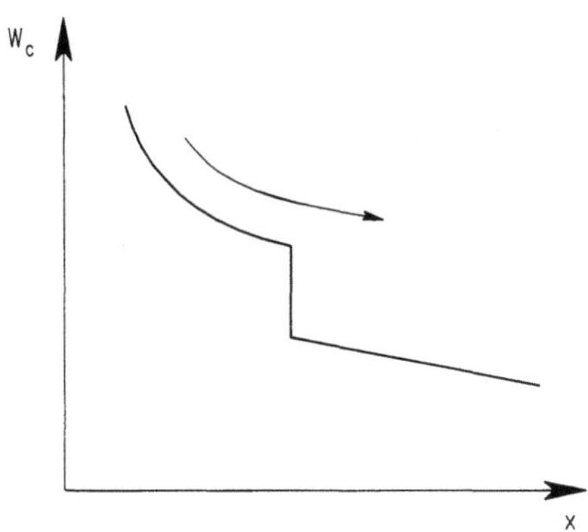

Fig. 7.1 Electron launcher

Table 7.1. *Semiconductor "midgap" energies W_B (in eV)*

Si	0.36
Ge	0.18
GaAs	0.50
AlAs	1.05
InAs	0.50
InP	0.76

$$\Delta W_c = W_{g2} - W_{g1} - W_{B2} + W_{B1}, \tag{7.1}$$

where W_B is a material parameter. Numerical values of W_B, calculated theoretically for a number of semiconductors, are taken from [31] and shown in Table 7.1. For alloys it is possible to estimate W_B by linear interpolation between the values in Table 7.1 according to the fractions of the different species. Taking the bandgap values of the involved semiconductors and the "midgap" energies from Table 7.1 and inserting into (7.1) yields the bandgap discontinuity with an accuracy of about 0.1 eV [31]

Following the emerging technology, the interest for simulation of heterojunction devices has recently grown considerably. While the carrier transport in gradual heterojunctions can be accounted for by appropriately adapted volume transport equations [10, 11], suitable interface conditions have to be set up at abrupt heterojunctions. In the present chapter, we are going to use the general methodology of Chapter 4 for the derivation of respective interface conditions at semiconductor heterojunctions. Before that, we will look at two examples of modern heterojunction devices, the Real-Space Transfer Transistor, and the Hetero-Bipolar Transistor (HBT), in order to obtain an impression of the (sometimes not so familiar) heterojunction devices to be simulated. After that, the inflow moments method will be used to derive interface conditions at a heterojunction for a hot-carrier transport model.

7.1 Heterojunction Devices

In this section, two examples of important and interesting heterojunction devices are briefly presented. These are the *Real-Space Transfer Transistor* and the *Hetero-Bipolar Transistor* (HBT). This collection does not cover by far the large variety of devices that is possible with heterojunction technology; for an overview see e.g. [142, 143, 144]. It is only intended as a particular sample in order to give a feeling of the subject. Especially, the large family of resonant tunneling devices [142] is not considered, because as pure quantum-effect devices they fall out of the scope of this book.

7.1.1 Real-Space Transfer Transistor

First, the real-space transfer effect, briefly mentioned in the introduction to this chapter, is considered in more detail. Figure 7.2 shows two semiconductor regions, separated by a conduction band edge discontinuity. In equilibrium (Fig. 7.2a), all electrons are in the deep valley of region 1. The potential step is so high that no electrons are able to cross the barrier. If the electrons are now heated by a strong electric field parallel to the interface, the electron temperature rises and consequently states of high energy become populated (Fig. 7.2b). Then, a considerable number of electrons have sufficient thermal kinetic energy to be able to cross the potential step and enter into region 2.

A particular transistor concept that uses the RST effect is the *charge injection transistor* (CHINT) [139, 145]. The basic structure of a CHINT is shown in Fig. 7.3. On a substrate of AlGaAs a thin layer of InGaAs is deposited, followed by a top layer made of GaAs. The InGaAs layer is *n*-doped and forms a conducting channel. Through heavily doped regions

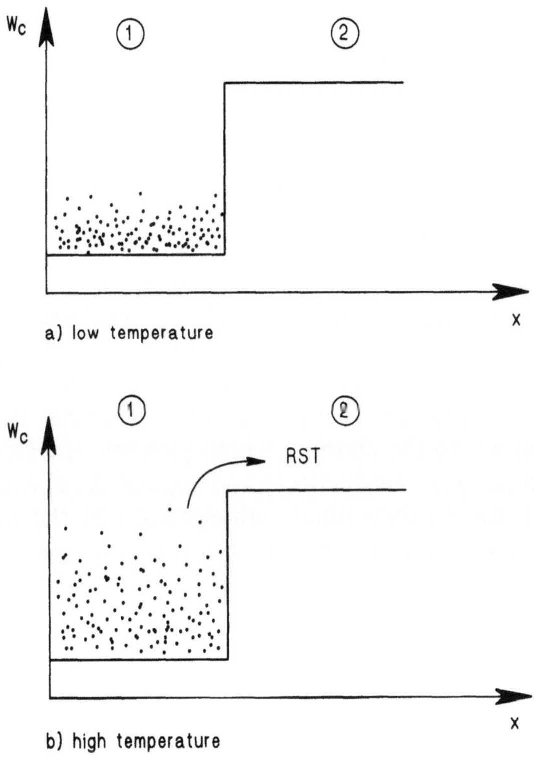

Fig. 7.2 The real-space transfer effect

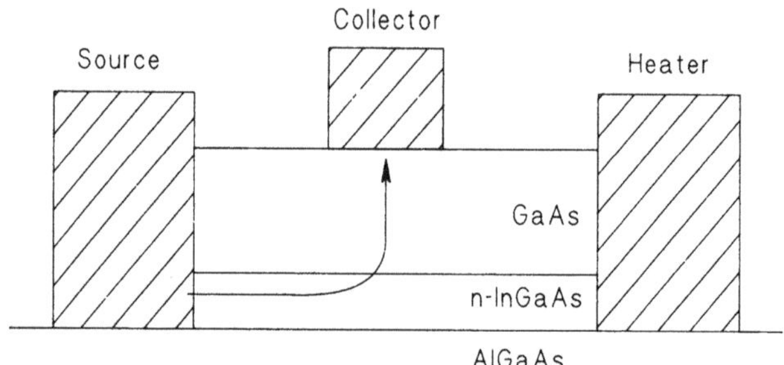

Fig. 7.3 Structure of the charge injection transistor

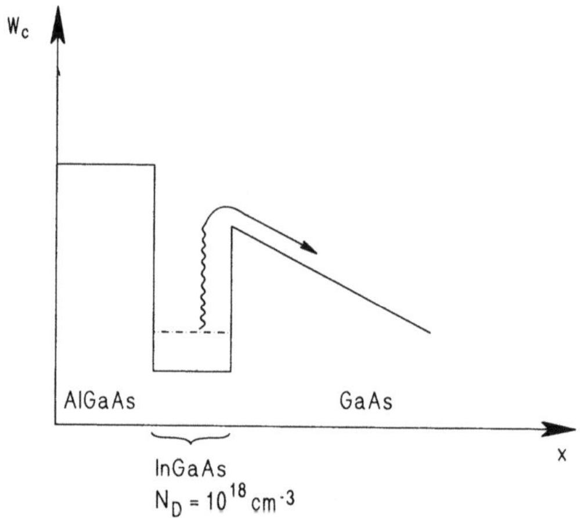

Fig. 7.4 Band diagram of the charge injection transistor

at the ends it is connected to the external contacts "source" and "heater". A third electrode "collector" is placed on top of the GaAs layer.

The band diagram of the CHINT on a vertical line through the collector is shown in Fig. 7.4. The AlGaAs substrate has the highest bandgap and thus forms a high barrier to the channel which prevents the electrons from entering the substrate. The conduction band edge of the channel is rather low and represents a steep valley, filled with electrons by the high doping. The channel is bounded by the potential step towards the GaAs layer, which is still large but smaller than the step on the substrate side.

A CHINT structure employing SiGe-Si heterojunctions in an alternative material combination has been reported in [146].

The functional principle of the CHINT is as follows. In equilibrium, all electrons are in the potential well. The collector is positively biased. Now a positive voltage is applied to the heater contact. Electrons start to drift through the channel towards the heater, but no current flows yet through the collector contact. At stronger electric fields in the channel, the electrons become heated. If the electron temperature gets sufficiently high, they surmount the barrier and are attracted by the positive collector, where they are absorbed. Thus, as the final outcome, the collector current is controlled by the heater voltage. Since the electron heating and cooling is very fast, the CHINT is suited as an amplifier for frequencies up to several 10 GHz. Another useful operation of the device is based on the following effect. If the heater voltage is raised outgoing from zero, the heater current rises until the RST sets in. Then a considerable fraction of the source current is transferred to the collector. Therefore, the heater current decreases, while the collector current starts to flow. Thus, in the heater-source circuit a negative differential resistance arises. In this operation mode the device is called *negative resistance field-effect transistor* (NERFET) [139, 145].

7.1.2 Hetero-Bipolar Transistor

The functional principle of the hetero-bipolar transistor (HBT) is identical to that of an ordinary bipolar transistor. However, the features of the transistor are improved by the inclusion of hetero-interfaces [147]. The basic structure of a Si-SiGe HBT is shown in Fig. 7.5. The emitter consists of highly-doped n-Si, the base of p-SiGe alloy, and the collector of n-Si again. The band diagram is depicted qualitatively in Fig. 7.6. Since SiGe has a smaller bandgap than silicon, band discontinuities are present at the emitter-base and base-collector junctions. These are step-like for the valence band, while the conduction band has small spikes.

The improvement of the HBT, compared to the ordinary bipolar transistor, is due to the additional energy barrier (ΔW_v in Fig. 7.6) for holes that are injected from the base into the emitter. These carriers contribute to the total emitter current but not to the collector current; thus they degrade the emitter efficiency and the transistor gain [19]. In the HBT, this part of the

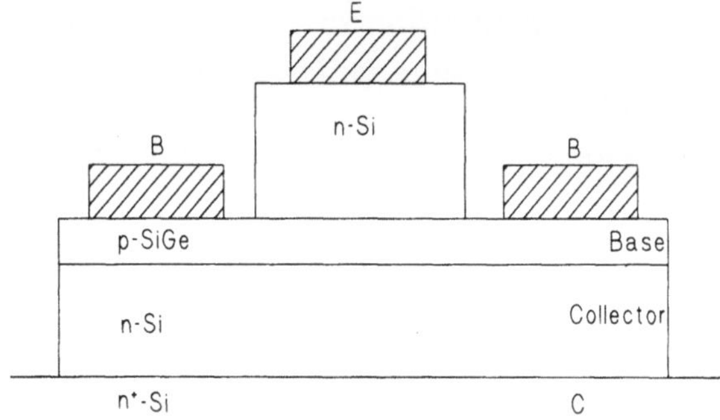

Fig. 7.5 Structure of the hetero-bipolar transistor

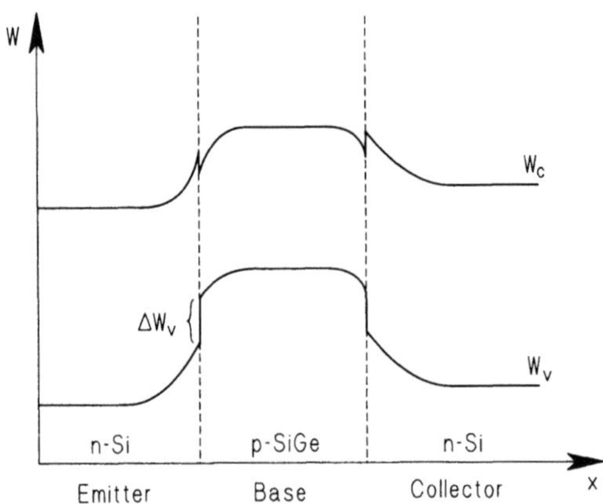

Fig. 7.6 Band diagram of the hetero-bipolar transistor

emitter current is suppressed by the valence band step at the emitter-base junction; as a result the emitter efficiency and hence the transistor gain increases [148]. Current gains of up to 5000 for a SiGe-Si HBT have been reported [149].

Alternatively, the doping concentration in the base of the SiGe-Si HBT can be increased compared to the Si-transistor without degrading the emitter efficiency at the same time. Thus, a thinner base and a lower base resistance is possible in the HBT, which results in a higher speed of transistor operation [147]. For instance, transit frequencies of more than 100 GHz for III-V HBTs [147] and 14 GHz for a SiGe-Si HBT [148] have been reported.

7.2 Interface Conditions for the Heterojunction

In this section, we derive interface conditions for electrons at a semiconductor heterojunction from our basic principles of Chapter 4. The presentation closely follows the first publication of the model in [73]; however, the derivation is presented with more emphasis on some of the details and approximations involved.

The basic configuration considered is the same as in Fig. 4.6. At the interface, two semiconductors 1 and 2 adjoin each other. The conduction band edge shows a discontinuity, characterized by ΔW_c. We assume again spherical parabolic bands in the two regions with effective masses m_1 and m_2.

It is clear from the preceding discussion that thermionic emission at the interface is the basic principle for the operation of the hetero devices presented in Section 7.1. It has been found from Monte Carlo simulations (including tunneling) of the CHINT [139] that the tunnel effect plays

only a negligible role for this device. Concerning the *npn*-HBT, tunneling appears to be important for the conduction band [150], while for the valence band it can be neglected because of the form of the band edge (see Fig. 7.6). Thus, we take the thermionic emission interface conditions presented in Section 4.3 and 4.5 as a starting point for the derivation. The inclusion of tunneling is straightforward along the lines presented in Chapter 4.
The general moment interface conditions for thermionic emission have been derived in Section 4.5 by the inflow moments method as (cf. (4.29) and (4.35))

$$\int_0^\infty dk_{2x} \int_{-\infty}^\infty dk_{2y}\, dk_{2z}\, v_{2x}(\mathbf{k}_2) M(\mathbf{k}_2) f_2(\mathbf{k}_2)$$

$$= \int_0^\infty dk_{2x} \int_{-\infty}^\infty dk_{2y}\, dk_{2z}\, v_{2x}(\mathbf{k}_2) M(\mathbf{k}_2) f_1(\mathbf{k}_1(\mathbf{k}_2)) \tag{7.2}$$

and

$$\int_{-\infty}^\infty dk_{1y}\, dk_{1z} \int_{-\infty}^{+\tilde{k}} dk_{1x}\, v_{1x}(\mathbf{k}_1) M(\mathbf{k}_1) f_1(\mathbf{k}_1)$$

$$= \int_{-\infty}^\infty dk_{1y}\, dk_{1z} \int_{-\infty}^{-\tilde{k}} dk_{1x}\, v_{1x}(\mathbf{k}_1) M(\mathbf{k}_1) f_2(\mathbf{k}_2(\mathbf{k}_1)), \tag{7.3}$$

where f_1 and f_2 are the electron distribution functions in regions 1 and 2, respectively. M determines the moment equation that the interface condition is associated to (see below). The mutual relations of $\mathbf{k}_1$ and $\mathbf{k}_2$ for this case according to (4.9) and (4.11) are

$$k_{2x}(\mathbf{k}_1) = \mathrm{sgn}(k_{1x}) \sqrt{\frac{m_2}{m_1}(k_{1x}^2 - \tilde{k}^2)}, \tag{7.4}$$

$$k_{1x}(\mathbf{k}_2) = \mathrm{sgn}(k_{2x}) \sqrt{\frac{m_1}{m_2} k_{2x}^2 + \tilde{k}^2}, \tag{7.5}$$

$$k_{1y} = k_{2y}, \qquad k_{1z} = k_{2z}, \tag{7.6}$$

with $\tilde{k}$ defined in (4.10).
In order to describe the real-space transfer effect, it is necessary that the interface condition takes an elevated electron temperature into account. Thus a hot electron transport model for the electron distribution in (7.2) and (7.3) has to be considered. The transport model we use for the following has been presented in [16] and [18]. It is based on the assumption

$$f_i(\mathbf{k}_i) = \exp\left(-\frac{\hbar^2 k_i^2}{2m_i k_B T_{ni}}\right) \left[1 - \frac{\tau_i}{k_B T_{ni}} \frac{\hbar \mathbf{k}_i}{m_i}(\mathbf{a}_i + k_i^2 \mathbf{b}_i)\right] c_i \tag{7.7}$$

for the electron distribution function. The symbol c_i is defined by

$$c_i = \exp\left(\frac{\phi_{ni} - W_{ci}}{k_B T_{ni}}\right) \tag{7.8}$$

as an abbreviation. The index $i = 1, 2$ labels the materials on both sides of the heterojunction. The assumption (7.7) results from an expansion of the antisymmetric part of a shifted Maxwellian distribution [12] up to third order in k. Alternatively, it can be obtained from a perturbative solution of the Boltzmann equation [16].

The unknowns of the assumption are the quasi-Fermi level $\phi_{ni}(\mathbf{x})$, the electron temperature $T_{ni}(\mathbf{x})$, and the parameters $\mathbf{a}_i(\mathbf{x})$ and $\mathbf{b}_i(\mathbf{x})$. The latter two terms are antisymmetric (odd in $\mathbf{k}$); they incorporate the flux property of the distribution since the symmetric part vanishes in the odd moments. Thus the flux densities $\mathbf{J}_{ni}$ and $\mathbf{Q}_{ni}$ of charge and energy are found as linear combinations of $\mathbf{a}_i(\mathbf{x})$ and $\mathbf{b}_i(\mathbf{x})$ by taking the first (2.10) resp. third (2.12) order moments of (7.7) [18],

$$\mathbf{J}_{ni} = \frac{q}{m_i} N_{ci} c_i \tau_i \left[\mathbf{a}_i + 5 \frac{m_i k_B T_{ni}}{\hbar^2} \mathbf{b}_i \right], \tag{7.9}$$

$$\mathbf{Q}_{ni} = -\frac{5}{2} \frac{k_B T_{ni}}{m_i} N_{ci} c_i \tau_i \left[\mathbf{a}_i + 7 \frac{m_i k_B T_{ni}}{\hbar^2} \mathbf{b}_i \right], \tag{7.10}$$

where N_{ci} is the effective state density of the semiconductor i as defined in (2.15).

The volume equations of the transport model can be derived by computing the first four moments of the Boltzmann equation with (7.7) inserted for the distribution function (see Chapter 2). The result of this operation consists of four partial differential equations for the four unknowns stated above. The particle conservation equation is obtained by setting

$$\text{particle conservation:} \quad M(\mathbf{k}_i) = 1 \tag{7.11}$$

when computing the moment of the Boltzmann equation. The energy conservation equation results when setting M to the kinetic energy $W(\mathbf{k})$:

$$\text{energy conservation:} \quad M = \frac{\hbar^2 k^2}{2m}. \tag{7.12}$$

The particle and energy flux equations are obtained from the moments where $M(\mathbf{k})$ according to (7.11) resp. (7.12) are multiplied by the group velocity $\mathbf{v}(\mathbf{k})$. This system of partial differential equations can be solved numerically in a device simulator to simulate the hot electron behaviour in a device. The solutions in the device regions separated by the heterojunction are connected by interface conditions, which we shall derive from (7.2) and (7.3) in the following.

The interface conditions for the transport model are obtained by inserting

(7.7) and (7.11) resp. (7.12) into the inflow moments (7.2) and (7.3). However, before doing so an approximation has to be made since not all of the resulting integrals are computable in closed form. Thereby, analytical formulas for the interface condition can be obtained.

For this purpose, we first split the distribution function f into a symmetric and an antisymmetric part with respect to $\mathbf{k}_i$:

$$f_i(\mathbf{k}_i) = f_{iS}(\mathbf{k}_i) + f_{iA}(\mathbf{k}_i). \tag{7.13}$$

The symmetric part f_{iS} originates from the first term in brackets in (7.7), while the antisymmetric part contains the **a**- and **b**-terms. The approximation to be introduced now consists of

a) dropping the antisymmetric part of f_1 on the right-hand side of (7.2), which then becomes

$$\int_0^\infty dk_{2x} \int_{-\infty}^\infty dk_{2y}\, dk_{2z}\, v_{2x}(\mathbf{k}_2)M(\mathbf{k}_2)f_2(\mathbf{k}_2)$$

$$= \int_0^\infty dk_{2x} \int_{-\infty}^\infty dk_{2y}\, dk_{2z}\, v_{2x}(\mathbf{k}_2)M(\mathbf{k}_2)f_{1S}(\mathbf{k}_1(\mathbf{k}_2)); \tag{7.14}$$

and

b) at the same time extending the k_{1x}-integration of f_{1A} on the left-hand side of (7.3) to infinity:

$$\int_{-\infty}^\infty dk_{1y}\, dk_{1z} \int_{-\infty}^{+\tilde{k}} dk_{1x}\, v_{1x}(\mathbf{k}_1)M(\mathbf{k}_1)f_{1S}(\mathbf{k}_1)$$

$$+ \int_{-\infty}^\infty dk_{1y}\, dk_{1z} \int_{-\infty}^\infty dk_{1x}\, v_{1x}(\mathbf{k}_1)M(\mathbf{k}_1)f_{1A}(\mathbf{k}_1)$$

$$= \int_{-\infty}^\infty dk_{1y}\, dk_{1z} \int_{-\infty}^{-\tilde{k}} dk_{1x}\, v_{1x}(\mathbf{k}_1)M(\mathbf{k}_1)f_2(\mathbf{k}_2(\mathbf{k}_1)). \tag{7.15}$$

In doing so, we neglect in (7.14) the fraction of the total flux entering region 2 from region 1 that is associated with f_{1A} (see the dashed vertical line in the upper left part of Fig. 4.11). Instead, in (7.15) we assumed that this flux portion is reflected in region 1. In other words, we "redirect" a small fraction of the flux that is originally crossing the interface from 1 to 2 by subtracting it from the transferred flux and adding it instead to the reflected flux. By this, the overall flux balance is maintained. It can be shown that the redirected flux portion is of the order $\exp(-\Delta W_c/(k_B T_{n1}))$ smaller than the total flux. If

$$\Delta W_c \gg k_B T_{n1} \tag{7.16}$$

holds, the factor is small, and the flux portion is negligible. Relation (7.16) is usually satisfied [145].

The integrals in (7.15) are more conveniently evaluated if the integration variable of the first integral on the left-hand side and that of the right-hand side integral is transformed to $\mathbf{k}_2$:

$$\int_{-\infty}^{\infty} dk_{2y}\, dk_{2z} \int_{-\infty}^{0} dk_{2x}\, v_{2x}(\mathbf{k}_2) M(\mathbf{k}_1(\mathbf{k}_2)) f_{1S}(\mathbf{k}_1(\mathbf{k}_2))$$

$$+ \int_{-\infty}^{\infty} dk_{1y}\, dk_{1z} \int_{-\infty}^{\infty} dk_{1x}\, v_{1x}(\mathbf{k}_1) M(\mathbf{k}_1) f_{1A}(\mathbf{k}_1)$$

$$= \int_{-\infty}^{\infty} dk_{2y}\, dk_{2z} \int_{-\infty}^{0} dk_{2x}\, v_{2x}(\mathbf{k}_2) M(\mathbf{k}_1(\mathbf{k}_2)) f_2(\mathbf{k}_2). \tag{7.17}$$

When transforming the first integral on the left-hand side of (7.15), the identity

$$\int_{-\infty}^{+\tilde{k}} dk_{1x}\, v_{1x} M f_{1S}$$

$$= \int_{-\infty}^{-\tilde{k}} dk_{1x}\, v_{1x} M f_{1S} + \underbrace{\int_{-\tilde{k}}^{+\tilde{k}} dk_{1x}\, v_{1x} M f_{1S}}_{= 0 \text{ since odd integrand}} \tag{7.18}$$

has been used. (Recall that according to the discussion in Section 4.4, M has to be an even function of k_{1x}.)

Now (7.7) and (7.11) resp. (7.12) can be inserted into (7.14) and (7.17) in order to obtain the interface conditions for the transport model. In this connection, the frequently occurring square of $\mathbf{k}_1(\mathbf{k}_2)$ evaluates to

$$(\mathbf{k}_1(\mathbf{k}_2))^2 = \frac{m_1}{m_2}(\mathbf{k}_2)^2 + \frac{2m_1}{\hbar^2} \Delta W_c \tag{7.19}$$

by (7.5), (7.6), and (4.10). (Note that (7.19) reflects again the conservation of energy at the interface.) By inspecting the integrals that result from this procedure, we find that the expressions contain only the following types of integrals,

$$I_v = \int_0^{\infty} dx\, x^v \exp(-x^2/2), \tag{7.20}$$

$$K_\mu = \int_{-\infty}^{\infty} dx\, x^\mu \exp(-x^2/2) \tag{7.21}$$

with v taking on the values 1, 2, 3, 4, 6 and μ the values 0, 2, 4, 6. These are readily evaluated to be

$$I_1 = 1, \qquad I_2 = \tfrac{1}{2}\sqrt{2\pi},$$
$$I_3 = 2, \qquad I_4 = \tfrac{3}{2}\sqrt{2\pi},$$
$$I_5 = 8, \qquad I_6 = \tfrac{15}{2}\sqrt{2\pi}, \tag{7.22}$$

and

$$K_0 = \sqrt{2\pi}, \qquad K_2 = \sqrt{2\pi},$$
$$K_4 = 3\sqrt{2\pi}, \qquad K_6 = 15\sqrt{2\pi}. \tag{7.23}$$

With the help of these integrals, the inflow moments can be evaluated by a straightforward but tedious calculation.

The interface conditions for the current continuity (i.e. particle or charge conservation) equations are obtained by setting $M = 1$ according to (7.11). Evaluation of (7.14) with $M = 1$ and (7.22)–(7.23) yields

$$(k_B T_{n2})^2 c_2 - \frac{\pi^2 \hbar^3}{q m_2} J_{2x} = (k_B T_{n1})^2 c_1 \exp\left(-\frac{\Delta W_c}{k_B T_{n1}}\right), \tag{7.24}$$

while the respective evaluation of (7.7) gives

$$-(k_B T_{n2})^2 c_2 - \frac{\pi^2 \hbar^3}{q m_2} J_{2x} = -(k_B T_{n1})^2 c_1 \exp\left(-\frac{\Delta W_c}{k_B T_{n1}}\right) - \frac{2\pi^2 \hbar^3}{q m_2} J_{1x}, \tag{7.25}$$

During the evaluation of (7.24) and (7.25), the arising terms containing **a** and **b** have been re-expressed by the charge and energy current densities using (7.9) and (7.10). In (7.25) the left and right-hand sides have been exchanged compared to (7.17).

Before discussing this result, we first evaluate the interface conditions for the energy conservation equation. Setting M according to (7.12) in (7.14) and computing the integrals following the discussion above results in

$$(k_B T_{n2})^3 c_2 + \frac{\pi^2 \hbar^3}{2 m_2} Q_{2x} = (k_B T_{n1})^3 c_1 \exp\left(-\frac{\Delta W_c}{k_B T_{n1}}\right). \tag{7.26}$$

The analogous evaluation of (7.17) for the reverse direction yields

$$-(k_B T_{n2})^3 c_2 \left(1 + \frac{\Delta W_c}{2 k_B T_{n2}}\right) + \frac{\pi^2 \hbar^3}{2 m_2} Q_{2x} - \frac{\pi^2 \hbar^3}{2 q m_2} \Delta W_c J_{2x}$$
$$= -(k_B T_{n1})^3 c_1 \left(1 + \frac{\Delta W_c}{2 k_B T_{n1}}\right) \exp\left(-\frac{\Delta W_c}{k_B T_{n1}}\right) + \frac{\pi^2 \hbar^3}{m_2} Q_{1x} \tag{7.27}$$

Again, $\mathbf{J}_n$ and $\mathbf{Q}$ have been substituted for the **a**- and **b**-terms, and the sides of Eq. (7.27) have been exchanged.

Eqs. (7.24)–(7.27) are in principle the interface conditions for the hot-electron transport model at an abrupt semiconductor heterojunction. However, for physical interpretation as well as numerical implementation, a rearrangement of the equation set is more suitable.

Adding together (7.24) and (7.25) gives (after cancelling a common factor) Eq. (7.28) below. Addition of (7.26) and (7.27) results in Eq. (7.29). While in the beginning the conditions have been set up by writing down the inflow into each region, in the last step we re-combined the flux fractions in order to obtain the total fluxes on each side of the interface.

Subtracting (7.25) from (7.24) and solving for J_{1x} yields Eq. (7.30), where (7.28) has been used. Finally, the analogous subtraction of (7.27) from (7.26) and solving for Q_{2x} yields Eq. (7.31) where we made use of (7.28) and (7.29).

$$J_{2x} = J_{1x}, \tag{7.28}$$

$$Q_{2x} = Q_{1x} + \frac{1}{q}\Delta W_c J_{2x}, \tag{7.29}$$

$$J_{2x} = q\left[v_{n2}(T_{n2})n_2 - \frac{m_2}{m_1}v_{n1}(T_{n1})n_1 \exp\left(-\frac{\Delta W_c}{k_B T_{n1}}\right)\right], \tag{7.30}$$

$$Q_{2x} = -2\left[k_B T_{n2} v_{n2}(T_{n2})n_2 \right.$$

$$\left. - \frac{m_2}{m_1}k_B T_{n1} v_{n1}(T_{n1})n_1 \exp\left(-\frac{\Delta W_c}{k_B T_{n1}}\right)\right]. \tag{7.31}$$

The electron concentrations n_1 and n_2 on both sides at the interface according to

$$n_i = N_{ci}c_i \tag{7.32}$$

have been used in (7.30–7.31), where N_{ci} is the effective state density of the respective region. The "emission velocity"

$$v_{ni}(T_{ni}) = \sqrt{\frac{2k_B T_{ni}}{\pi m_i}} \tag{7.33}$$

has been introduced as an abbreviation.

The final interface conditions for thermionic emission of hot electrons at a semiconductor heterojunction thus consist of Eq. (7.28), (7.29), (7.30), and (7.31). They represent four conditions connecting the four unknowns ϕ_n, T_n, $\mathbf{J}_n$, and $\mathbf{Q}$ on both sides of the interface. Besides these interface conditions for the balance equations, we still have the conditions for the Poisson equation as discussed in Chapter 3.

The interface conditions found can be interpreted as follows. Eqs. (7.28) and (7.29) reflect again the principle of flux balancing at the heterojunction, which has been the basis of the derivation from the beginning. Equation

(7.28) simply states particle conservation at the interface. Equation (7.29) relates the energy and current fluxes on both sides of the heterojunction. In (7.29), we recognize the Peltier resp. Seebeck effect, i.e. the transformation of current density into energy flow and vice versa at abrupt heterointerfaces [8]. Intuitively, the second term on the right-hand side of (7.29) reflects the conversion of potential energy into kinetic energy when the electron falls down the step ΔW_c. It can be shown that the flux balancing equations come out independently of the specific assumption for the distribution function. Eq. (7.30) describes the thermionic emission across the energy step by relating the interface current to the electron concentrations and thus to the quasi-Fermi levels on both sides of the interface. The current can be viewed as the net current of two fluxes in opposing directions. Each partial flux is expressed as the product of the interface concentration n_i and the emission velocity v_{ni}, which characterizes the emission efficiency of the carriers in region i across the junction. The exponential factor of n_1 in (7.30) and (7.31) expresses the fact that only the fraction of the total electron concentration on the region 1 side that is energetically above the barrier is able to cross the potential step.

Eq. (7.31) represents the energy current density according to thermionic emission at the heterointerface. As the electric charge current density, it is expressed as the difference of two opposing fluxes. Interestingly, the same emission velocities as in (7.30) occur. The local thermal energy $k_B T_n$ appears as an additional factor and represents the thermal energy that is transported by the thermionic emission.

Sometimes it is convenient to write the emission equations (7.30) and (7.31) in a different way. By algebraic transformations we get

$$J_{2x} = q[v_{n2}(T_{n2})n_2 - v_{n2}(T_{n1})n_0] \tag{7.34}$$

$$Q_{2x} = -2[k_B T_{n2} v_{n2}(T_{n2})n_2 - k_B T_{n1} v_{n2}(T_{n1})n_0], \tag{7.35}$$

where n_0 is defined by

$$n_0 = \left(\frac{T_{n1}}{T_{n2}}\right)^{3/2} N_{c2} \exp\left(\frac{\phi_{n1} - W_{c2}}{k_B T_{n1}}\right). \tag{7.36}$$

This formulation is more similar to the customary description of thermionic emission in Schottky barriers (compare with Eq. (6.168) in Section 6.3 on Schottky contacts). However, in contrast to (6.168) we have now different emission velocities because of the different electron temperatures in the two regions. The concentration n_0 is analogous to (6.170); it could be interpreted as the fraction of the total electron concentration on the region 1 side that is energetically above the barrier, and thus is able to participate in the thermionic emission process.

If we compare the expression for the emission velocity (7.33) with the corresponding expression (6.171) of Chapter 6, we note that they differ by a factor of two, although both appear in an identical manner in the current equations. The same distinction can be found frequently when comparing emission velocities of different publications in the literature. It can be shown that the factor arises from the use of different distribution functions in the derivation of the models. The use of the double-semi-Maxwellian distribution of Section 6.2.2 regularly yields an emission velocity that is half as large as the value one obtains from using the volume distribution. However, both of the mentioned distribution functions have their disadvantages, which have already been discussed in Section 6.2.2; hence it is difficult to give preference to one over the other. A more consistent modelling can be expected by a thorough investigation of the transition from the interface to the volume distribution, maybe along the lines pointed out in [113] and [123] for the metal-semiconductor contact (see Section 6.3).

7.3 Literature Review

The advent of heterojunction devices, and the corresponding wish to simulate them, has raised the necessity of interface conditions that connect a device simulator's solutions in each semiconductor region at the hetero-interface. Thus, since the end of the 1970's the literature on interface conditions at heterojunctions has been considerably growing.
In 1979, Wu and Yang [151] have proposed an interface condition for the drift-diffusion transport model. The model was based on thermionic emission at the heterojunction, but assumed a constant electron temperature throughout the device.
Maeda [68] considered electron transport across heterojunctions from the Boltzmann equation point of view. His treatment is equivalent to the derivation in Section 4.2.
In the following years, a sequence of similar interface conditions for the simulation of electron transport across heterojunctions have been presented in [152, 153, 154]. All these papers considered only the case of isothermal thermionic emission, i.e. the electron temperature was assumed to be equal to the constant lattice temperature. The interface conditions were essentially identical to (7.28), (7.30) with $T_{n1} = T_{n2}$.
First treatments of heterojunctions for hot-electron transport models have been reported by Mosko et al. [155, 156] and myself [20]. Mosko considered the special case of energy transport under the condition of vanishing particle current across the interface. In [20], I proposed conditions which stated continuity of quasi-Fermi level and electron temperature at the interface instead of (7.30)–(7.31). The same conditions have been used in simulations of a charge injection transistor [157]. However, as noted in [153] (at least for the isothermal case), the continuous quasi-Fermi level condition leads to unsatisfactory results when compared with a full consid-

eration of thermionic emission. In 1992 [73], the improved model of Section 7.2 was published, where I pointed out that the continuity of quasi-Fermi level and electron temperature is only valid for the case of small particle and energy fluxes across the interface. In fact, we can see that Eqs. (7.30) and (7.31) tend to $\phi_{n1} = \phi_{n2}$ and $T_{n1} = T_{n2}$ only in this limit.

In 1991, Tait and Westgate [158] published a paper on a one-dimensional drift-diffusion simulation of heterostructures. The simulated devices contained a ramp structure, i.e. a linear increase of the conduction band edge over a certain region which was followed by an abrupt drop. The ramp, surrounded by GaAs regions, was fabricated by linearly grading the aluminum content of AlGaAs from 0 to 0.6 using molecular beam epitaxy. Besides the usual interface conditions for the Poisson equation (3.11) and (3.12) the authors used a nonlinear current density equation which couples the bulk regions on each side as an interface condition of the electron continuity equation. In the present notation, their expression for the current density at the interface reads

$$J_{2x} = \frac{q}{2} v_{n2}(T) n_2 \left[1 - \exp\left(\frac{\phi_{n1} - \phi_{n2}}{k_B T}\right) \right], \qquad (7.37)$$

where ϕ_{n1} and ϕ_{n2} are the quasi-Fermi energies in regions 1 and 2, respectively. Eq. (7.37) is fully equivalent to (7.30) in the isothermal ($T_{n1} = T_{n2} = T$) case, as can be shown by transforming (7.30) into (7.37) using the relation between n and ϕ_n.

At first glance, one might be surprised that the interface condition appears to be independent of the potential step ΔW_c of the heterojunction at all. However, this is actually not the case; while in (7.30) the interface current can be computed if n_1 and n_2 are known, the evaluation of the current in (7.37) requires the computation of the Fermi level difference from the concentrations with the help of

$$\exp\left(\frac{\phi_{n1} - \phi_{n2}}{k_B T}\right) = \frac{n_1}{n_2}\left(\frac{m_2}{m_1}\right)^{3/2} \exp\left(-\frac{\Delta W_c}{k_B T}\right), \qquad (7.38)$$

which results from (2.14). This is where ΔW_c enters the model again when implementing the interface condition into a numerical scheme.

From the equivalence of (7.37) and (7.30), it is clear that (7.37) is in principle a thermionic emission model. Tunneling has been included *a posteriori* in a very coarse way. It has been assumed that the tunnel effect lowers the barrier ΔW_c by an amount W_T, given by the simple formula

$$W_T = \begin{cases} -\dfrac{dW_c}{dx} x_T, & \dfrac{dW_c}{dx} < 0, \\ 0, & \text{otherwise.} \end{cases} \qquad (7.39)$$

The barrier lowering was assumed to be proportional to the band edge gradient, where the tunneling length x_T was assumed to be a given fixed constant.

A simulation of a heterojunction device that included tunneling at the abrupt heterointerface more consistently has been reported by Yang et al. in 1993 [159]. The treatment is similar to the concepts of Chapters 4 and 6. In the one-dimensional bulk regions, the conventional drift-diffusion equations have been solved, while the thermionic emission and tunneling effects are placed into the interface condition at the heterojunction. This interface condition determined the relation between the (distinct) quasi-Fermi levels on both sides. As in Section 7.2, two interface conditions have been used. Although not explicitly stated as such, the first one used is identical to (7.28), i.e. particle conservation. It is implicitly incorporated in the numerical scheme described in [159]. The second condition accounts for the current transport across the heterojunction. Transcribed to the present notation, it reads

$$J_{2x} = \frac{q}{2}(1 + \delta)\left[v_{n2}(T)n_2 - \frac{m_2}{m_1}v_{n1}(T)n_1 \exp\left(-\frac{\Delta W_c}{k_B T}\right)\right]. \qquad (7.40)$$

This equation should be compared to (7.30). However, because only drift-diffusion is considered in [159], hot-electron effects cannot be accounted for, and thus $T_{n1} = T_{n2} = T$ has to be set for the comparison. In this case, both equations are identical, except for the aforementioned factor of 2 in the emission velocity and a factor $(1 + \delta)$. The parameter δ accounts for the tunneling effect. It is defined by

$$\delta = \frac{1}{k_B T}\exp\left(\frac{W_{c2}}{k_B T}\right)\int_{W_{min}}^{W_{c2}} dW\, D(W)\exp\left(-\frac{W}{k_B T}\right), \qquad (7.41)$$

where $D(W)$ is the WKB tunneling probability as given in Eq. (6.12), and W_{c2} is the conduction band edge on the region 2 side of the interface. The lower bound of the integral is defined as $W_{min} = \max[W_{c1}(x_0), W_{c2}(w)]$, which is the minimum energy from where tunneling is at all possible; w denotes the edge of the depletion region on side 2.—Expression (7.40) has been used before by Grinberg et al. [160] in an analytical investigation of hetero-bipolar transistors.

It can be shown that Yang's current formula would come identically out of (6.26) if we insert non-degenerate hemi-Maxwellians for f_1 and f_2, and if the inflow into region 2 would be computed at the location x_m where $W_{c2}(x_m) = W_{min}$. However, this point may lie appreciably apart from the interface if band bending is small in region 2. It is thus somewhat inconsistent that the current computed in this way is then (numerically) injected into region 2 immediately at the interface; computation of the current at the end of the tunneling range would be preferrable.

In contrast to our treatment of tunneling at the metal-semiconductor contact in Section 6.2.1, Yang et al. did not approximate the tunneling probability by an analytical formula but computed it numerically from the actual band profile. While this approach works in one dimension, it is presently not clear whether it is still feasible in the more complex situation of a two- or three-dimensional simulation.

Hjelmgren [128] considered hot electrons in device simulations at a Schottky contact. According to the general discussion in Section 2.5, we need in this case two boundary conditions instead of the four interface conditions (7.28)–(7.31) of the heterojunction interface. Only inflow into region 2, i.e. Eqs. (7.24) and (7.26), now has to be taken into account; the other two conditions have to be discarded. As a result, we arrive at Eqs. (7.34)–(7.35) as the boundary conditions for the hot-electron Schottky contact. Exactly these conditions have been introduced by Hjelmgren with less formal arguments.

8 MOSFET Gate

Although the gate of a MOSFET (metal-oxide field-effect transistor) is in fact a semiconductor-insulator interface, which has already been discussed in Chapter 5, it shall receive special attention in this chapter because of the special importance of MOSFETs. In this context, we will not discuss the entirety of electron transport along the channel and across the gate; this topic would fill a whole book on its own, see e.g. [4]. Merely, we shall concentrate on peculiar effects on carrier transport that are caused by the presence of the insulator interface.

8.1 Surface Scattering

In an enhancement-mode MOSFET [19] the conducting path between the source and drain contacts is maintained by inverting the channel by a correspondingly large voltage on the gate. The gate is separated from the channel by a thin insulating layer. This voltage creates a large electric field in the channel normal to the insulator surface. Minority carriers are strongly driven by the field towards the interface, where they assemble to form the conducting inversion layer. Applying a voltage between the source and drain contacts makes the carriers start moving towards the drain along the channel and parallel to the interface; hence a current flows between source and drain.

During their flow along the channel, the carriers are frequently driven towards the insulator interface by the strong field caused by the gate voltage. Hence, frequent collisions with the interface occur, where the carriers impinge on the insulator potential wall and are reflected back into the channel.

Because of the situation in a MOSFET channel just described—transport parallel to the surface, strong field normal to the surface—scattering at the insulator surface plays an important role. Normally, this scattering hampers the current flow along the interface. As a result, the effect of surface scattering can be expressed by a reduction of mobility in the vicinity of the surface. Realistic device simulation of MOSFETs has to take this effect into account.

This phenomenon is experimentally investigated by measuring the drain current under a low source-drain voltage, and varying the voltage at the gate [161]. In this way, the conditions along the gate stay nearly homogeneous, while the strength of the normal electric field is controlled by the gate. Thus, the force that drives the carriers against the interface and the corresponding amount of surface scattering can be varied and the effects studied. The drain current and the gate voltage, respectively, can be evaluated to obtain the mobility and the normal electric field in the channel. However, the conditions in the direction normal to the surface are strongly inhomogeneous. Hence, the measurements can give only an "effective mobility" and an "effective field" [162]. These effective quantities are averages of the corresponding local quantities over the total channel depth. The dependence of the effective mobility on the effective field is reported to be insensitive to the bulk doping and the source-substrate voltage [162]; this is the so-called "universal mobility curve".

Drift-diffusion device simulators need the mobility as a transport parameter in order to compute the local drift current from the local electric field. However, as Wong [162] pointed out, the effective mobility cannot be successfully used for this purpose, at least in two- or three-dimensional device simulation. This is the case because the effective mobility is a *global* average, given as a function of another global quantity, i.e. the effective field, while a device simulator operates with *local* quantities. Thus it is necessary to describe surface scattering on a more microscopic level in order to obtain the required local mobility.

In this context, one has to regard the imperfection of the insulator interface (see Fig. 8.1). In case of an ideal interface, the electrons are reflected under

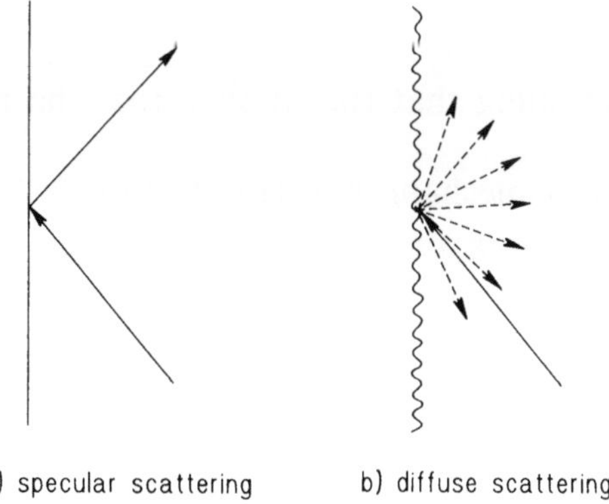

Fig. 8.1 Specular and diffuse scattering

conservation of energy and transverse momentum. Because this results in simply turning back the normal momentum component, the angle of incidence equals the angle of reflection, just like a beam of light reflecting from a mirror. For this reason, one speaks of "specular scattering" (Fig. 8.1a). With specular scattering, the velocity components parallel to the interface are unaffected by the scattering at the surface; hence mobility degradation does not take place in this case. However, in reality several mechanisms like surface roughness scattering, scattering by interface charges, or scattering by surface phonons are in effect that change the tangential momentum as well. An incident electron is scattered into a reflected direction at random, according to the scattering probability. This mechanism is called "diffuse reflection" (Fig. 8.1b). Diffuse reflection thus creates corresponding degradations of the tangential flow near the interface [14].

As Selberherr notes [48, 163], these mechanisms have not been investigated as deeply as one would expect. Ando et al. [164] presented a quantum-mechanical approach to surface scattering, but this analysis was restricted to very low temperatures and is thus not directly applicable to devices at room temperature. On the level of the Boltzmann equation, one usually disregards the microscopic scattering mechanisms at the surface, and phenomenologically assumes a diffuse reflection with equal probabilities for all possible reflection directions [165]. In drift-diffusion simulators, empirical mobility models are used that reduce the drift velocity in the vicinity of the surface.

In the following, we analyze surface scattering in the general framework of interface transport as introduced in Chapter 4. Interface conditions for the Boltzmann equation are introduced. The next section will show how a surface mobility model can be derived from these conditions for a simplified case.

We start with the interface conditions of Section 4.7. The specular scattering is given by the scattering kernel

$$R_{\mathrm{spec}}(\mathbf{k}', \mathbf{k}) = \delta(k_x + k_x') \, \delta(k_y - k_y') \, \delta(k_z - k_z'), \tag{8.1}$$

which results from (4.82) by taking the $\mathbf{k}$-$\mathbf{k}'$-relationship according to the spherical band, recalling that transition across the gate insulator is neglected.

The diffuse scattering mechanism is described by

$$R_{\mathrm{diff}}(\mathbf{k}', \mathbf{k}) = \frac{v_x(k_x)}{\pi k^2 v(k)} \delta(k - k'), \tag{8.2}$$

where the delta function takes care that the scattering is elastic. The prefactor has been determined such that the normalization condition (4.75) as well as the reciprocity condition—i.e. detailed balance in equilibrium [14]—is satisfied. In [14] a different formulation of diffuse reflection for gas streaming along a surface was presented (which can be traced back to Maxwell), which assumes that the reflected particles have an equilibrium distribution. This means that the mechanism is inelastic, because hot particles are cooled to the temperature of the surface during the reflection (accommodation). For surface scattering in MOS channels, however, no experimental evidence for such cooling is present. Hence, for the following the formulation (8.2) is preferred.

Both the elastic diffuse scattering above and the accommodating diffuse reflection in [14] constitute only phenomenological descriptions of the reflection at an imperfect surface; microscopic investigations of the interaction between the particle and the surface are not included. Also, established microscopic models are presently not available at all, and the literature is still controversy. Tang et al. [166] reported measurements of the interface roughness by means of x-ray diffraction technique for different types of Si-SiO$_2$ layers. They found roughness amplitudes of 0.38 nm for a native oxide, 0.2 nm for a thermally grown oxide, and 0.46 nm for a chemically-grown RCA-clean type of oxide. The authors suggested to make correlations of the measured roughness to degradation of surface transport, which however has not yet been done with their data. While in [167] investigations have been reported indicating that the surface roughness scattering should be too weak to be responsible for the mobility degradation, a correlation between the microscopically observable interface roughness and the surface mobility in MOSFETs has been found experimentally, reported in [168].

A possibility to account for the surface quality at least phenomenologically is to assume that the surface is in part specularly reflecting and in part diffusively reflecting. For this purpose a parameter α (the "accommodation coefficient" [14]) is introduced. The parameter may vary from 0 to 1 and "mixes" the scattering of the perfect surface (8.1) with the scattering of the rough surface (8.2):

$$R(\mathbf{k}, \mathbf{k}') = (1 - \alpha)R_{\text{spec}}(\mathbf{k}', \mathbf{k}) + \alpha R_{\text{diff}}(\mathbf{k}', \mathbf{k}). \tag{8.3}$$

In a sense, α describes the imperfection of the surface: If $\alpha = 0$, the surface is perfect, $\alpha = 1$ describes a rough surface, and α between 0 and 1 describes a surface somewhere inbetween these two extremes. The surface scattering model (8.3) dates back to Maxwell, too.

A resonable value of α for Si-MOSFETs seems to be very small. Selberherr et al. [163] found from numerical calculations with a model potential and comparison with experimental data that $\alpha < 0.2$ is required. Fiegna and Sangiorgi [169] made comparisons of Monte Carlo simulations and experiments and found a value of $\alpha = 0.085$. Hence it can be expected that the diffuse surface scattering is very effective in reducing the mobility near the insulator interface.

8.2 Surface Mobility Degradation

Now we turn to the question of how to include the surface scattering mechanism into device simulations. In Section 4.7, an interface condition for the distribution function has been stated, where the energy and the transverse momentum at the interface are not necessarily conserved. Although our elastic diffuse reflection model (8.2) conserves energy, the momentum of the reflected electron changes in general. Hence, the non-conservative boundary condition has to be used here.
We assume that a transition of the electron into the insulator does not take place. Thus we obtain from (4.77) with $D = 0$ the boundary condition for the Boltzmann equation at a rough insulator surface

$$v_x(\mathbf{k})f(x_0, \mathbf{k}) = -\int_{k'_x < 0} d^3\mathbf{k}' \, v_x(\mathbf{k}')R(\mathbf{k}', \mathbf{k})f(x_0, \mathbf{k}'), \qquad (8.4)$$

with R from (8.3). As usual, this boundary condition is valid for inflow $(v_x(\mathbf{k}) > 0)$. By insertion of (8.2) one easily verifies that the diffuse scattering is indeed isotropic, since $v_x(\mathbf{k})$ cancels on both sides of the equation.
For device simulations that directly solve the Boltzmann equation, (8.4) is all that we need. Translated into the language of Monte Carlo the condition says that a particle impinging on the surface is reflected back into the device according to the probability distribution $R(\mathbf{k}', \mathbf{k})$. In fact, device simulators that solve the Boltzmann equation with the Monte Carlo method make use of a boundary condition like this [7, 165].
The situation is more complicated in the case of approximate transport models like drift-diffusion or spherical harmonics expansion. In principle, boundary conditions for these transport models have to be derived again by the procedure introduced in Section 4.4, i.e. taking the inflow moments of the boundary condition (8.4).
The boundary condition for the drift-diffusion transport model is obtained by taking the zero order inflow moment of (8.4). With the help of the normalization condition (4.75) we easily find from this moment the boundary condition

$$J_{nx} = 0. \qquad (8.5)$$

However, this same condition was found in Chapter 5, Eq. (5.38) when we considered the semiconductor-insulator interface without taking surface scattering into account. Thus, the drift-diffusion approximation is not able to carry more information than telling us that the current across the insulator boundary vanishes; the effect of surface scattering drops out in the approximation.
The reason for this behaviour is that the effect of surface scattering on current flow parallel to the interface does not take place at the surface

alone, but also in the bulk within a certain distance to the surface. It has been pointed out by Ohno [170] that the mobility degradation near the MOS surface can be interpreted by considering shear stress in the flow, which is due to the viscosity of the electron gas. The reasoning is as follows. The roughness of the surface absorbs momentum when electrons are reflected (cf. the discussion above). Thus a force acts on the surface, which is balanced by a shear stress in the gas flow at the surface. This stress reduces the tangential velocity of the flow at the surface. The viscosity of the electron gas then "transports" the surface shear stress into the bulk, because a transversal velocity difference generates shear stress between adjacent streams to compromise the velocity difference [170]. As a net result, the surface roughness produces a velocity reduction at the interface, which decays into the bulk because of the viscosity of the electron gas.

It can be shown that viscosity is a higher order effect, which is not included in the first order approximation of drift-diffusion transport. Also in the spherical harmonics expansion of the Boltzmann equation [171] it is not covered by the expansion using only the first two terms. For this reason, it is difficult to include surface roughness self-consistently into *boundary* conditions for these transport models. Hence, surface scattering cannot be contained in the boundary condition (8.5), but has to be modelled by a modification of the transport in the volume.

In the following paragraphs a simple model after Schrieffer [172] and Greene [173] is presented, which illustrates the way how this problem can be attacked. Since we have seen that the effect is not described within first order approximations, we have to investigate the transport near the surface on the level of the Boltzmann equation. Having found a solution for the distribution function near the surface, we will be able to derive a consistent approximation in the drift-diffusion frame by taking again the respective moment.

We consider a cross section of a MOSFET as depicted in Fig. 8.2. A positive gate voltage is applied that tilts the bands towards the insulator interface and creates an electric field E_x in the x-direction normal to the insula-

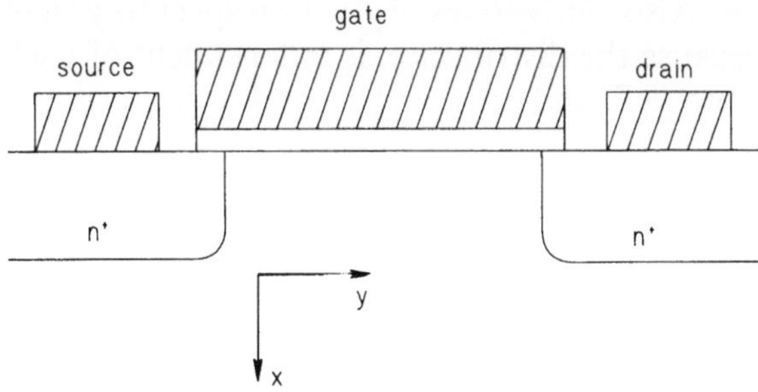

Fig. 8.2 MOSFET cross section

tor interface. If zero voltage is applied to the source and drain contacts, no current flows and the electrons are in equilibrium, described by a Maxwell-Boltzmann distribution $f_0(x, \mathbf{k})$. The electron distribution is independent of y. The insulator interface is described by the boundary condition (8.4) with (8.3), giving

$$
\begin{aligned}
v_x(k_x)f(x_0, \mathbf{k}) = & -\alpha \frac{\hbar k_x}{\pi m_n k^3} \int_{k'_x < 0} d^3\mathbf{k}' \, \delta(k - k')k'_x f(x_0, \mathbf{k}') \\
& - (1 - \alpha) \int_{k'_x < 0} d^3\mathbf{k}' \, \delta(k_x + k'_x) \, \delta(k_y - k'_y) \\
& \times \delta(k_z - k'_z) \, v_x(\mathbf{k}')f(x_0, \mathbf{k}').
\end{aligned}
\tag{8.6}
$$

In order to have a tractable problem, we use a simplified Boltzmann equation, i.e. the relaxation time approximation of the Boltzmann equation [4, 80]

$$
\frac{\hbar \mathbf{k}}{m_n} \operatorname{grad} f - \frac{q\mathbf{E}}{\hbar} \operatorname{grad}_k f = \frac{f_0 - f}{\tau}.
\tag{8.7}
$$

The relaxation time τ is assumed to be constant. In equilibrium, the electric field only has a component in the x-direction. During the integration of the Boltzmann equation, we will ignore the variation of E_x with x for simplicity [172]. Treatments of the problem that get by without this assumption can be found in [172, 173, 80, 174, 4]. Here the emphasis is put on the illustration how the boundary condition enters the derivation; for this purpose the simpler formulation is sufficient.

We now turn on a small electric field E'_y in the y-direction by applying a small voltage to the drain. This field perturbs the equilibrium distribution, and we want to approximate the resulting distribution function with the perturbation method. Thus we make the assumption

$$
f = f_0 + f_1 \qquad (f_1 \ll f_0)
\tag{8.8}
$$

for the distribution function, and we will keep only terms that are linear in the perturbation. Also, derivatives of f with respect to y shall be considered of first order because the distribution is independent of y when $E'_y = 0$. Inserting (8.8) into (8.7) and dropping second order terms, we get

$$
\frac{\hbar k_x}{m_n} \frac{\partial f_1}{\partial x} - \frac{qE_x}{\hbar} \frac{\partial f_1}{\partial k_x} + \frac{f_1}{\tau} = -\frac{q\hbar}{m_n k_B T} k_y E'_y f_0,
\tag{8.9}
$$

where we accounted explicitly for the electric field components occurring. This partial differential equation can be solved by making a variable transformation from x to the newly defined variable

$$v = -\frac{W_c(x)}{k_B T} - \frac{\hbar^2 k_x^2}{2m_n k_B T}. \tag{8.10}$$

Note that $\tilde{f}_0$ in the new variables depends only on (v, k_y, k_z). The original dependence on x and k_x is absorbed in v. The perturbation f_1 transforms according to

$$f_1(x, \mathbf{k}) = \tilde{f}_1\left(-\frac{W_c(x)}{k_B T} - \frac{\hbar^2 k_x^2}{2m_n k_B T}, \mathbf{k}\right) = \tilde{f}_1(v, \mathbf{k}), \tag{8.11}$$

and (8.9) written in terms of $\tilde{f}_1$ reads

$$-\frac{qE_x}{\hbar}\frac{\partial \tilde{f}_1}{\partial k_x} + \frac{\tilde{f}_1}{\tau} = -\frac{q\hbar}{m_n k_B T} k_y E_y' \tilde{f}_0. \tag{8.12}$$

This equation is now a first order differential equation with respect to the variable k_x alone; its general solution for constant E_x is [75]

$$\tilde{f}_1(v, \mathbf{k}) = -\frac{q\tau\hbar}{m_n k_B T} k_y E_y' \tilde{f}_0 + C(v, k_y, k_z)\exp\left(\frac{\hbar k_x}{q\tau E_x}\right), \tag{8.13}$$

where the integration constant C still may depend on the other variables of (8.12). Returning to the original variables and inserting f_1 into (8.8), we get the solution for f

$$f(x, \mathbf{k}) = f_0(x, \mathbf{k}) - \frac{q\tau\hbar}{m_n k_B T} k_y E_y' f_0(x, \mathbf{k})$$

$$+ C\left(-\frac{W_c(x)}{k_B T} - \frac{\hbar^2 k_x^2}{2m_n k_B T}, k_y, k_z\right)\exp\left(\frac{\hbar k_x}{q\tau E_x}\right). \tag{8.14}$$

Note that the first two terms in (8.14) constitute the solution of the Boltzmann equation in the semiconductor bulk. The first order moment (2.10) of these terms gives the familiar expression (2.21) for the current density, with the bulk mobility (cf. Eq. (2.20))

$$\mu_B = \frac{q\tau}{m_n}. \tag{8.15}$$

The third term is responsible for the mobility degradation near the surface. The integration constant C has to be determined from the boundary conditions of (8.7). For C we make the ansatz

$$C(v, k_y, k_z) = k_y E_y' \tilde{C}(v, k_y, k_z), \tag{8.16}$$

because the y-symmetry of the problem is broken only by the perturbation $k_y E_y'$, and since we keep only first order terms, a linear dependence on $k_y E_y'$ is to be expected.

According to Section 2.5, the boundary condition for the Boltzmann equation is the prescription of the distribution function on the inflow boundaries. Regarding the present problem, one boundary is at the insulator interface $x = x_0$ with the inflow boundary condition (8.6). The second boundary is at $x \to \infty$. Since we do not have any external inflow at large distances from the interface, the second boundary can be considered by allowing only solutions that tend to the bulk distribution in the limit of large x.

Hence we have to insert (8.14) with (8.16) into (8.6) and solve for the function $C(v, k_y, k_z)$. The boundary condition thus has to be seen as a function of v, to be taken at the same time at $x = x_0$. However, according to (8.10), the prescription of x and v fixes the value of k_x. Thus, in order to obtain $C(v, k_y, k_z)$, the value of k_x has to be set to

$$k_{xs}(v) = \sqrt{\frac{2m_n}{\hbar^2}[-vk_BT - W_c(x_0)]} \tag{8.17}$$

in the boundary condition.

Inserting (8.14) into (8.6), we note in the first place that the equilibrium distribution f_0 satisfies the boundary condition by itself; hence only the terms containing f_1 remain in the equation. For the first integral on the right-hand side of (8.6), we find

$$\int_{k_x' < 0} d^3\mathbf{k}'\, \delta(k - k')k_x' f_1(x_0, \mathbf{k}') = 0, \tag{8.18}$$

because the integrand is odd in k_y and is integrated over a symmetric k_y-interval. The second integral can be easily evaluated because of the δ-functions. Eq. (8.6) thus becomes

$$\frac{\hbar k_x}{m_n} f_1(\mathbf{x}_0, \mathbf{k}) = (1 - \alpha)\frac{\hbar k_x}{m_n} f_1(\mathbf{x}_0, -k_x, k_y, k_z). \tag{8.19}$$

This boundary condition has been described essentially already in 1938 by Fuchs [175]. Note that the inflow part of f_1 at the boundary turns to zero in the case of full diffuse scattering ($\alpha = 1$).

Insertion of f_1 from (8.14) and (8.19) yields

$$\alpha \frac{q\tau\hbar}{m_n k_B T} f_0(x_0, k_{xs}(v), k_y, k_z)$$

$$= \tilde{C}(v, k_y, k_z)\left[\exp\left(\frac{\hbar k_{xs}(v)}{q\tau E_x}\right) - (1 - \alpha)\exp\left(-\frac{\hbar k_{xs}(v)}{q\tau E_x}\right)\right], \tag{8.20}$$

where we used the fact that f_0 is even in k_x. Eq. (8.20) can be solved for $\tilde{C}$, and inserting everything into (8.14), we obtain the final solution

$$f(x, \mathbf{k}) = f_0(x, \mathbf{k}) - \frac{q\tau\hbar}{m_n k_B T} k_y E_y' f_0(x, \mathbf{k})$$

$$\times \left[1 - \alpha \frac{\exp\left(\dfrac{\hbar}{q\tau E_x}[k_x - k_{xs}(x, k_x)]\right)}{1 - (1 - \alpha)\exp\left(-\dfrac{2\hbar}{q\tau E_x} k_{xs}(x, k_x)\right)} \right], \qquad (8.21)$$

where $k_{xs}(x, k_x)$ is (8.17) with v inserted from (8.10). Note that $k_{xs}(x, k_x)$ gets large as x moves into the bulk and $W_c(x)$ significantly exceeds $W_c(x_0)$. Comparing (8.21) with (8.14), we note that the difference to the bulk solution consists of the second term inside the brackets. It vanishes towards the bulk with increasing $W_c(x)$ and decreasing E_x as required. Seen in a different view, the expression in the brackets can be considered as a factor that diminishes the bulk relaxation time in the vicinity of the surface by reason of the surface scattering.

Having found a solution on the level of the distribution function, we are now able to project the result into approximate transport descriptions in order to obtain consistent models of surface scattering on that level. It should be noted at this point that in [172, 173, 80, 174] the *effective* mobility was determined from the distribution function by averaging over the channel thickness, while Hänsch [4] calculated the actual *local* mobility from the solution.

The description for the drift-diffusion transport model is obtained by taking the respective moments of (8.21). The result can be interpreted as an increase of the scattering rate resp. a decrease of the mobility in the vicinity of the surface. This surface effect then decays towards the bulk according to the exponential function in (8.21).

In the following, this assertion is demonstrated with an approximate solution for the drift-diffusion transport model. The drift-diffusion transport model is derived from the Boltzmann equation by computing the zero and first order moments.

The zero order moment of (8.21) is unaffected by the surface term since only the equilibrium term contributes to that moment. Hence, the continuity equation (2.17) remains unchanged.

Taking the first order moment with respect to k_y gives the current density in the y-direction,

$$J_{ny}(x) = A(x) \cdot [I_0 - I_s] I_2 I_0, \qquad (8.22)$$

where I_0 and I_2 are defined by

$$I_n = \int_{-\infty}^{\infty} dk \, k^n \exp(-ak^2), \qquad (8.23)$$

I_s by

$$I_s = \alpha \int_{-\infty}^{\infty} dk_x \frac{\exp(-ak_x^2 + bk_x - b\sqrt{k_x^2 + c})}{1 - (1 - \alpha)\exp(-2b\sqrt{k_x^2 + c})}, \tag{8.24}$$

and $A(x)$ collects the remaining factors independent of **k**. The constants in (8.23) and (8.24) are defined by

$$a = \frac{\hbar^2}{2m_n k_B T}, \qquad b = \frac{\hbar}{q\tau E_x}, \qquad c = \frac{2m_n}{\hbar^2}[W_c(x) - W_c(x_0)]. \tag{8.25}$$

I_0 turns out to be $\sqrt{\pi/a}$, while I_2 cancels below when considering the surface mobility.

Since I_s vanishes in the bulk of the semiconductor, we obtain from (8.22) for the current density in the bulk

$$J_{ny}^{(\text{bulk})} = A(x) \cdot I_0 I_2 I_0 = q\mu_B n E_y'. \tag{8.26}$$

Inserting (8.26) into (8.22), we get

$$J_{ny}(x) = q\mu_B \left[1 - \frac{I_s}{I_0} \right] n E_y'. \tag{8.27}$$

Now it becomes clear that the surface scattering effect can be interpreted as decreasing the bulk mobility μ_B by the factor $(1 - I_s/I_0)$ to give a position dependent reduced mobility $\mu_n(x)$ in the vicinity of the surface:

$$\mu_n(x) - \mu_B \left[1 - \frac{I_s}{I_0} \right]. \tag{8.28}$$

Eq. (8.28) could be evaluated for a specific problem by numerical integration of I_s. Instead, we like to proceed with a further approximation that enables us to obtain an analytical result for the surface mobility which reflects nicely the observed behaviour.

The evaluation is carried out in Appendix D. We finally obtain the degraded mobility near the interface as given by

$$\boxed{\begin{aligned} \mu_n(x) = \mu_B \frac{1}{2} \Bigg[&2 - \alpha + \alpha \operatorname{erf}\left(\sqrt{u(x)}\right) \\ &- \alpha \exp\left(\frac{E_s}{E_x}\left[\frac{E_s}{E_x} - 2\sqrt{u(x)}\right]\right) \cdot \operatorname{erfc}\left(\frac{E_s}{E_x} - \sqrt{u(x)}\right) \\ &- F(u(x), E_x, \alpha) \Bigg], \end{aligned}} \tag{8.29}$$

where

$$u(x) = \frac{W_c(x) - W_c(x_0)}{k_B T} \tag{8.30}$$

and

$$E_s = \sqrt{\frac{m_n k_B T}{2q^2 \tau^2}} = \frac{1}{\mu_B}\sqrt{\frac{k_B T}{2m_n}} \tag{8.31}$$

have been used as abbreviations, and the function F is defined in the Appendix. This mobility corresponds to the surface mobility μ^{LIS} as introduced by Selberherr [48], which could be incorporated into a complete mobility model along the lines proposed by that author.

It is clear that the quantitative applicability of (8.29) is limited because of the assumptions of constant E_x and τ. However, the structure of (8.29) may serve as a template for building model expressions for the surface mobility, with parameters to be determined from experiment. Fitting parameters can be α and eventually a factor in front of of E_s, which may partially compensate the error made when assuming a constant E_x. In this sense, we may relax our initial assumption of constant E_x by allowing E_x in (8.29) to be the *local* normal electric field component.

Fig. 8.3 shows the mobility degradation according to (8.29) as a function of distance from the surface for an analytical approximation [176] of the typical potential and field in a MOSFET channel. A value of $\alpha = 0.1$ has been taken. As expected, we find a severe mobility degradation in the vicinity of the surface. The mobility approaches the bulk value for distances from the surface larger than the mean free path of the electrons. The curve in Fig. 8.3 is similar in shape to the various empirical functions for surface mobility modelling that have been discussed by Selberherr et al. [163].

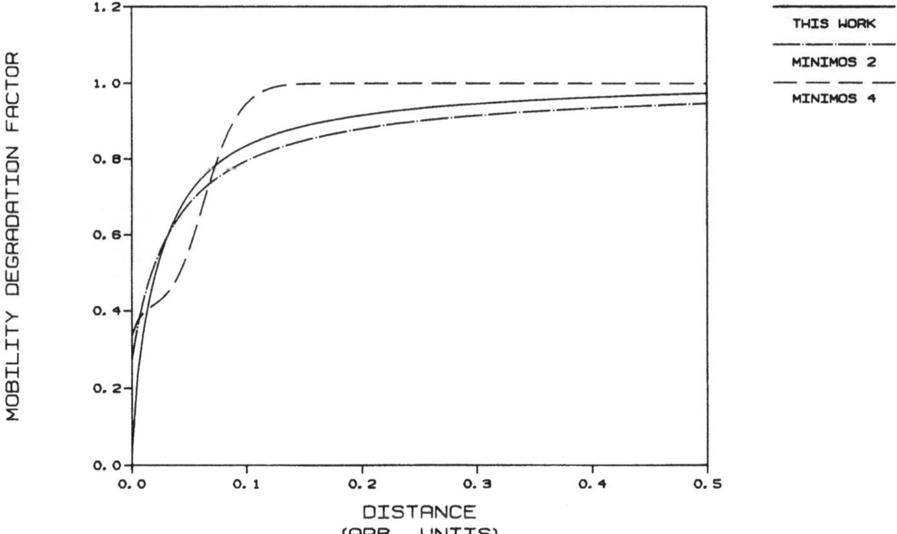

Fig. 8.3 Mobility degradation as a function of distance from the surface (arbitrary units)

For comparison, the corresponding curves of the MINIMOS 2 and the MINIMOS 4 mobility models are shown in Fig. 8.3. The model presented here comes very close to the empirical MINIMOS 2-model, while the agreement with MINIMOS 4 is still quite reasonable.

Hänsch [4] followed an approach for deriving a surface mobility similar to the present one. He solved the Boltzmann equation (8.12) for arbitrary $E_x(x)$ dependence (resembling the treatment in [80, 174]), and assumed in the following an exponential shape for $E_x(x)$ when evaluating the current density (cf. (8.27)). He then numerically computed the remaining integrals and replaced them by empirical fitting functions in order to arrive at a surface mobility model, suited for the use in a device simulator.

Discretization 9

9.1 Volume Discretization

Carrier transport has been described in the previous chapters in terms of
fields, i.e. in terms of physical quantities that are defined continuously for
all points in space (and time) in the simulation domain. In the drift-diffusion transport model, for instance, the state of a semiconductor device is
described by the electron and hole concentration fields and the potential
field. It is the task of device simulation to solve the transport equations
according to the input of geometry, doping, and boundary conditions; the
primary result of the simulation is the distribution of the fields inside the
device. Additional information, like for instance the terminal currents, can
be computed from the fields in a post-processing step.

The infinite spatial and temporal resolution of the continuous fields requires the discretization of the transport equations for solving them numerically on a finite digital computer. In principle, the discretization serves to
create an approximation of the continuous problem into a problem that
contains only a discrete, finite number of variables.

To this end, in principle one has to set up approximations for the continuous fields that are characterized by a finite number of variables. After that,
the continuous transport equations have to be transformed into a corresponding set of equations that operate on the discrete fields. The numerical
solution of these discrete transport equations then yields the discrete field
representation, which in turn can be transformed back to give an approximation of the continuous fields as the result of the simulation.

Normally, the field discretization is done by splitting the simulation domain into a number of small elements, and assigning a field value to each
element. By this, the continuous field is approximated by a vector of field
values, associated with the elements. An interesting alternative to this procedure has been proposed by Axelrad [177], who approximated the fields
by expanding them into a Fourier series of a finite number of terms. However, the majority of device simulators use the method of volume element
discretization. Thus, we shall limit the discussion to this type.

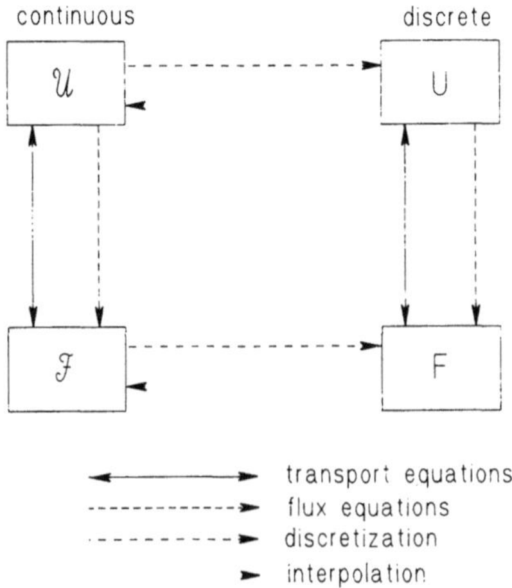

Fig. 9.1 Continuous and discrete field representations

The relations between the continous and discrete representations are schematically depicted in Fig. 9.1 [178]. The state of the semiconductor device
is described by a set of conserved quantities, the unknowns $\mathcal{U}$. In the balance equations, these are the even moments of the distribution function.
Associated with each variable is a corresponding flux density, which are the
odd moments (those constructed with $\mathbf{v}M$) of the distribution function.
In addition, we have the electrostatic potential as another state variable,
with the electric displacement $\mathbf{D}$ as the corresponding flux. (For a discussion of thermodynamic state and flux variables in semiconductors, see also
[179].)
The fluxes $\mathcal{F}$ depend on the unknowns $\mathcal{U}$ through the flux equations or
"constitutive equations", as indicated in Fig. 9.1. The flux equations are
obtained from the odd moments of the Boltzmann equation. The transport
equations itself are the conservation equations, which are the result of the
even moments of the Boltzmann equation. They are in general of the form

$$\operatorname{div} \mathcal{F}(\mathbf{x}) = \mathcal{S}(x), \tag{9.1}$$

i.e. the divergence of the flux equals some source term $\mathcal{S}$, which in turn
may depend on the unknowns $\mathcal{U}$. (For the potential, the analogous equation is the third Maxwell equation (2.26), which is also of the form (9.1)).
Every discretization scheme in principle has to establish the relationships
between the continuous and discrete quantities as shown in Fig. 9.1 and
derive from these the discrete representation of the transport equations.
The following discussion has the three-dimensional case in mind; the simplification to lower dimensions is straightforward.
The transport equations can be discretized with the "finite difference
method" [48], which relies on a rectangular grid and substitution of the
differential operators with difference operators. However, the discretization

method most commonly used in device simulation is the "box integration" scheme (see e.g. [180, 181, 48, 133, 182]). The finite box method can be viewed as a generalization of finite differences to more irregular grids; in the case of rectangular grids the discrete equations of both methods turn out to be the same. Because of its importance, we shall concentrate on the finite box method, making occasionally some remarks on finite differences. A third alternative is the "finite element method" [181], which however has not gained much importance in semiconductor device simulation.

In the box integration method, the simulation domain is partitioned into small elements, called "boxes". The situation is depicted in Fig. 9.2. For ease of drawing, a simple rectangular partitioning of the domain is shown; however, the method is also valid for more irregularly shaped elements. In particular, the Voronoi cell method, also known as "Dirichlet tesselation", has recently received much attention in the device simulation community (e.g. [183, 184, 185, 182]). This method is a very convenient and well-suited way to discretize an irregularly shaped simulation domain.

A particular point inside each box is chosen, which is the "node" associated with the respective box. From a numerical point of view, the center of gravity of the box is preferred as the node. This requirement, however, is not so very strict in normal cases, as will be shown below. The nodes can be connected by lines to form a "grid" or "mesh". Practical grid generators usually start with placing the nodes according to the geometry, and determine the elements once the mesh is completed [184].

The discrete transport conservation is obtained from (9.1) by integrating the equation over one element i,

$$\int_{V_i} d^3\mathbf{x} \operatorname{div} \mathscr{F}(\mathbf{x}) = \int_{V_i} d^3\mathbf{x} \mathscr{S}(\mathbf{x}), \tag{9.2}$$

where V_i is the volume of the element. Applying Gauss' law to the left-hand side yields an integral over the total surface of the element

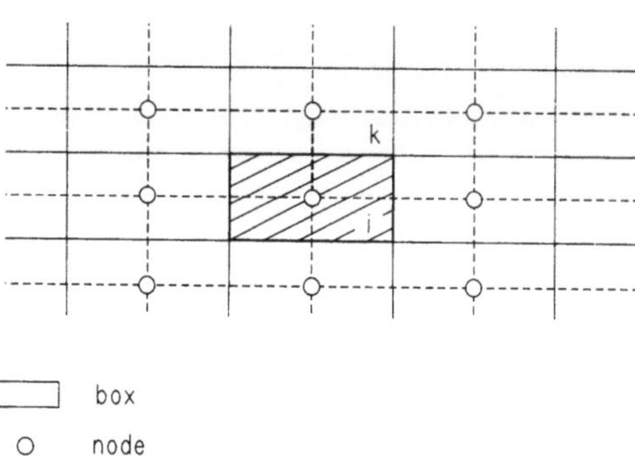

Fig. 9.2 Box integration scheme

$$\int_{\partial V_i} d\mathbf{A} \cdot \mathscr{F}(\mathbf{x}) = \int_{V_i} d^3\mathbf{x}\, \mathscr{S}(\mathbf{x}). \tag{9.3}$$

Now the right-hand side is approximated by the discrete source value S_i, multiplied by the volume V_i of the element. Considering the left-hand side, we note that according to Fig. 9.2 the total surface of the element i consists of a number of plane sections (faces) that connect the element i with its surrounding neighbours k. The surface integral over each of these faces is approximated by the discrete normal flux on the plane F_{ik}, multiplied by the area A_{ik} of the face:

$$\sum_k A_{ik} F_{ik} = V_i S_i. \tag{9.4}$$

Here k runs over those elements that are neighbours to element i.
Division by V_i yields the final form of the discrete conservation equation

$$\sum_k \frac{1}{h_{ik}} F_{ik} = S_i \tag{9.5}$$

with $1/h_{ik} = A_{ik}/V_i$. The operator acting on F_{ik} in (9.5) can be seen as a discrete version of the divergence operator (compare Eqs. (9.1) and (9.5)).
The discrete flux values F_{ik} are approximated by differences of quantities associated with the nodes. These approximations can be obtained by projecting the flux equation onto the connecting line between nodes i and k and (approximately) solving the originating ordinary differential equation (see e.g. [133]).
For a consistent discretization it is absolutely necessary that the approximation of the flux on the faces of the element has an accuracy with an error of $O(h^2)$, where h is a measure of the grid size (see above). Only if this requirement is satisfied, a Taylor expansion of (9.5) reproduces the continuous equation (9.1) in the limit of vanishing grid size, because then the $O(1)$-terms of F_{ik} in (9.5) (i.e. the fluxes itself) cancel, and the $O(h)$-terms combine to give the divergence of the flux. As a result of the division by h_{ik}, the left-hand side of (9.5) is of order $O(1)$, with an error of $O(h)$. The importance of this point has also been stressed in [48, 180].
Usually, S_i is computed by evaluating $\mathscr{S}$ at the node x_i, which means that the source term is assumed to be constant in the element. This assumption is correct up to an error of order $O(h^2)$ if the node is at the center of gravity of the box; otherwise the error is only $O(h)$. As a consequence it can be expected that the approximation of the right-hand side, when going from (9.3) to (9.4), becomes less accurate the more the node is distant from the center of gravity, if $\mathscr{S}$ is inhomogeneous. However, since the error of the left-hand side is $O(h)$, this is not crucial as long as the derivative of $\mathscr{S}$ is not $O(h^{-1})$ or larger
In the end we obtain a discrete version (9.5) of the conservation equation where the fluxes and sources are expressed as functions of the discrete

variables U_i. By assembling Eqs. (9.5) for all elements i, we obtain an equation system for the values U_i at all the elements, whose solution basically constitutes the result of the simulation.

So far we have considered only elements in the interior of the simulation domain. Elements that are adjacent to a boundary or interface have to be treated specifically because of the boundary resp. interface conditions.

While most authors in the device simulation literature put much emphasis on the description of the utilized volume discretization, they describe in many cases only roughly the discretization and implementation of the interface and boundary conditions they used. Hence, a clear picture of the used type of discretization unfortunately often cannot be obtained.

9.2 Boundary Conditions

The usual boundary configuration is depicted in Fig. 9.3. The difference to Fig. 9.2 is that at least one of the faces of the element lies on the boundary of the simulation domain. This means that a neighbour is missing for the respective face, and the corresponding flux term in (9.5) does not exist. It is at this place where the boundary condition comes into play.

For the numerical treatment of the boundary, i.e. the incorporation of the boundary condition into the discrete equation system, we have to distinguish what kind of continuous quantities are given on the boundary: the unknown $\mathcal{U}$ (Dirichlet boundary condition, BD), or the normal flux $\mathcal{F}$ through the boundary surface (Neumann boundary condition, BHN or BIN). The basic types of boundary conditions are displayed in

$$\mathcal{U} = f(\mathcal{U}, \mathcal{G}) \quad \text{(BD)}, \tag{9.6}$$

$$\mathcal{F} = 0 \quad \text{(BHN)}, \tag{9.7}$$

$$\mathcal{F} = f(\mathcal{U}, \mathcal{G}) \quad \text{(BIN)}, \tag{9.8}$$

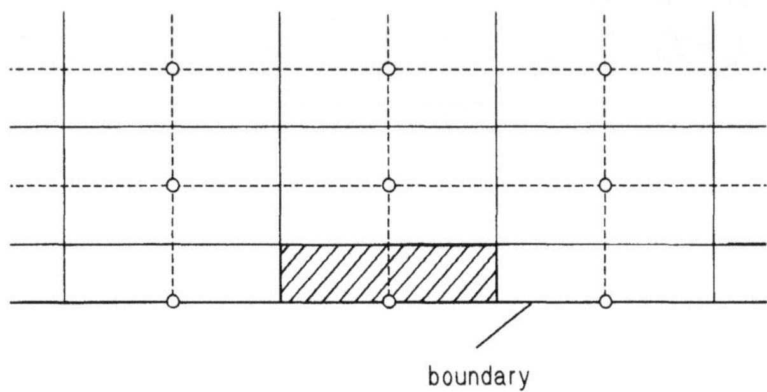

Fig. 9.3 Discretization element located at a boundary

where $\mathscr{G}$ stands symbolically for something that is "given", i.e. a boundary parameter or external stimulation; the symbol $\mathscr{U}$ denotes one or several of the state variables, hence a mixing of different variables is not explicitly shown. Also possible dependencies of boundary parameters on $\mathscr{U}$ or $\mathscr{F}$ (e.g. a dependence of the Schottky barrier height on the electric field) are disregarded in above notation in order to emphasize the basic relations.

In case of a *Dirichlet boundary condition*, the discrete value U_i of the unknown is directly given by the boundary condition. Hence, in these cases the conservation equation (9.5) is not considered, but is *replaced* by the boundary condition specifying U_i.

Of the boundary conditions we have encountered so far in the previous chapters, these are (see also Tables 9.1 and 9.2)

- conditions (5.35) or (5.36) for the continuity equation at the semiconductor-insulator interface (strong recombination);

Table 9.1. *Dirichlet boundary conditions (part 1)*

Semiconductor-insulator interface	
$np = n_i^2$	(5.35)
$\phi_n = \phi_p$	(5.36)
Non-ideal contact	
$W_c(x_T) = \phi_M + \phi_B - W_T^{(D)}$	(6.125)
$W_c(x_T) = \phi_M + \phi_B + W_T^{(D)}$	(6.132)
$W_c(x_T) = W_N(\phi_n, \phi_p)$	(6.126), (6.133)
$\phi_p = \phi_M$	(6.110)

Table 9.2. *Dirichlet boundary conditions (part 2)*

Schottky contact	
$W_c(x_T) = \phi_M + \phi_B$	(6.163)
$\phi_p = \phi_M$	(6.175)
Ohmic contact	
$n - p - N_D + N_A = 0$	(6.182)
$W_c = \phi_M - k_B T \, \mathrm{Arsinh}\left(\dfrac{N_D - N_A}{2n_{ie}}\right) + \dfrac{1}{2}W_g + \dfrac{k_B T}{2}\ln\left(\dfrac{N_c \gamma_n}{N_v \gamma_p}\right)$	(6.186)
$\phi_n = \phi_M$	(6.184)
$\phi_p = \phi_M$	(6.185)

- conditions (6.125) and (6.132) for the Poisson equation at the non-ideal contact (barrier lowering according to tunneling);
- conditions (6.126) and (6.133) for the Poisson equation at the non-ideal contact (charge neutrality);
- approximate condition (6.110) for the minority carriers at the non-ideal contact (small minority current);
- condition (6.163) for the Poisson equation at the Schottky contact (Schottky barrier height);
- condition (6.175) for minority carriers at the Schottky contact (small minority current);
- conditions (6.182) and (6.186) for the Poisson equation at the ohmic contact (charge neutrality);
- conditions (6.184–6.185) for the continuity equations at the ohmic contact (strong tunneling).

It is worth noting that we found Dirichlet boundary conditions almost entirely at metal-semiconductor contacts (except the strong recombination case at the insulator interface). This reflects the physical fact that the metallic contact allows to directly control the state of the semiconductor at the interface from the external world by making use of metallic wires or leads. Some of the contact boundary conditions (Eqs. (6.110), (6.163), (6.175), (6.184), (6.185)) listed above depend only on the Fermi energy ϕ_M of the metal, which can be controlled by external circuitry. These are the most simple cases; just the externally defined Fermi level has to be inserted into the boundary conditions for computing the value of the unknown variable on the boundary node. Because these conditions are independent of the solution inside the semiconductor, they introduce an element of stability into the numerical solution process.

The Poisson boundary conditions stating charge neutrality at the boundary (Eqs. (6.126, 6.133, 6.182)) involve the Fermi energies of electrons and holes as well as the band edge (resp. the electrostatic potential). Since this introduces a coupling of the Poisson and continuity equations at the boundary, convergence of the numerical scheme is more difficult to achieve. In the case of an *ideal* ohmic contact, however, where both the electron and hole quasi-Fermi levels are equal to that in the metal (cf. (6.184) and (6.185)), the coupling drops out of the equation; see (6.186). Hence, also the ideal ohmic contact model has a stabilizing effect on the numerical solution. The numerical implementation of these conditions is straightforward, since they involve only local quantities, which can be taken from the actual node under consideration.

The numerically most challenging cases are the Poisson boundary conditions in depletion (Eqs. (6.125) and (6.132)) of the non-ideal contact model, because they involve the band edge, the metal Fermi level as well as the majority carrier Fermi level and the normal electric field. The majority

Fermi level and the electric field enter the equation through the formula for the tunneling energy range. For these boundary conditions, a discrete approximation for the electric field at the boundary node has to be computed, which makes the boundary condition dependent on the values of neighbouring nodes, too. For this reason, this kind of boundary condition is no longer of pure Dirichlet type.

Finally, the strong recombination boundary condition at the insulator interface involves both electron and hole quasi-Fermi levels and thus couples the two continuity equations. The implementation of this condition is nevertheless straightforward.

The *Neumann boundary conditions* involve derivatives of the unknowns at the boundary[1]. We distinguish homogeneous (Eq. (9.7)) and inhomogeneous Neumann boundary conditions (Eq. (9.8)). In the homogeneous case, the flux and hence the derivative normal to the boundary is zero; inhomogeneous Neumann boundary conditions describe physical flux mechanisms at the interface. In case of a Neumann boundary condition, the discrete conservation equation (9.5) still has to be solved; however, the flux across the boundary face (cf. Fig. 9.3) is computed from the boundary condition.

The node of the boundary element is usually located on the boundary, not in the interior of the element as in the bulk. Since the boundary condition specifies the flux directly on the boundary, the accuracy requirement is no problem when inserting the flux into (9.5). On the other hand, in a *finite difference* discretization the flux approximation on the boundary requires special attention in this configuration in order to keep the discretization error to $O(h^2)$ [48].

Sometimes, one finds the notion that the inhomogeneous Neumann condition (BIN) is replaced by (BHN), where the right-hand side of (9.8) is transformed into an equivalent volume source term. This is possible according to (9.5) by a respective division of $f(U, G)$ by h.

Neumann boundary conditions have occurred in the previous chapters in the following equations (see also Tables 9.3–9.5):

- condition (5.38) or (5.39) for the majority current density, and (5.28) for the total current density at the semiconductor-insulator interface (totally insulating interface);
- conditions (5.27) and (5.32) for the continuity equations at the semiconductor-insulator interface (surface recombination);
- condition (5.41) for the Poisson equation at the semiconductor-insulator interface (surface charge, no external field);

---

[1] Somewhat relaxing the mathematical terminology, we use the term Neumann boundary condition if the condition is specifying the flux, instead of just the derivative, of $\mathcal{U}$.

Table 9.3. *Neumann boundary conditions (part 1)*

Semiconductor-insulator interface	
$J_{nx} = 0$	(5.38), (8.5)
$J_{px} = 0$	(5.39)
$J_{nx} + J_{px} = 0$	(5.28)
$J_{px} = qs_p \dfrac{np - n_i^2}{n + n_1 + \dfrac{s_p}{s_n}(p + p_1)}$	(5.27)
$J_{px} = qs_p(p - p_0)$	(5.32)
$\dfrac{\partial \varphi_s}{\partial x} = \dfrac{\sigma}{\varepsilon_s}$	(5.41)
$\dfrac{\partial \varphi_s}{\partial x} = 0$	(5.42)

Table 9.4. *Neumann boundary conditions (part 2)*

Non-ideal contact	
$J_{nx} = -\dfrac{qm_n(k_B T)^2}{2\pi^2 \hbar^3}[j_T(\psi_B) - j_T(\psi_B - \psi_c)]$	(6.130)
$J_{px} = -\dfrac{qm_p(k_B T)^2}{2\pi^2 \hbar^3}\exp(-\psi_{Bp} - \eta_T)[\exp(\psi_{cp}) - 1]$	(6.131)
$J_{nx} = \dfrac{qm_n(k_B T)^2}{2\pi^2 \hbar^3}\exp(-\psi_B - \eta_T)[\exp(\psi_c) - 1]$	(6.137)
$J_{px} = \dfrac{qm_p(k_B T)^2}{2\pi^2 \hbar^3}[j_T(\psi_{Bp}) - j_T(\psi_{Bp} - \psi_{cp})]$	(6.138)

Table 9.5. *Neumann boundary conditions (part 3)*

Schottky contact	
$J_{nx} = \dfrac{qm_n(k_B T)^2}{2\pi^2 \hbar^3}\exp(-\psi_B)[\exp(\psi_c) - 1]$	(6.166)
$J_{px} = -\dfrac{qm_p(k_B T)^2}{2\pi^2 \hbar^3}\exp(-\psi_{Bp})[\exp(\psi_{cp}) - 1]$	(6.174)
Ohmic contact	
$J_{nx} = \dfrac{1}{q\rho_c}(\phi_n - \phi_M)$	(6.183)

- condition (5.42) for the Poisson equation at the semiconductor-insulator interface (neither surface charge nor external field);
- conditions (6.130), (6.131), (6.137), (6.138) for the continuity equations at the non-ideal contact (thermionic and tunneling emission currents);
- conditions (6.166) and (6.174) for the continuity equations at the Schottky contact (thermionic emission currents);
- condition (6.183) for the majority carrier continuity equation at the resistive ohmic contact (ohmic resistance);
- condition (8.5) for the continuity equation at the rough MOSFET gate interface.

Homogeneous Neumann conditions in this list are (5.38), (5.39), (5.34), (5.42), and (8.5). They can be considered in the box integration scheme simply by omitting the boundary surface when assembling the flux contributions in (9.5) [181]. For finite differences a method called "mirror imaging" is discussed by Selberherr [48], which accounts for keeping the discretization error to $O(h^2)$ in rectangular elements.

Concerning the inhomogeneous Neumann conditions, the flux through the boundary surface has to be computed according to the boundary condition and inserted into the flux balance (9.5). It is also possible to use a modification of the mirror imaging method for inhomogeneous conditions by adding the inhomogenity term to the mirror flux [48].

The current conditions (5.32) at the insulator interface, (6.166) and (6.174) at the Schottky contact, and (6.183) at the resistive ohmic contact involve not only the gradient of the Fermi level, but also the respective Fermi level itself; in the Schottky contact case even in a nonlinear manner. Hence, these conditions do not represent Neumann conditions in the pure sense; sometimes they are referred to as "boundary conditions of the third kind". As a consequence, a respective entry on the diagonal of the Jacobi matrix of the discrete equation system appears. However, the conditions involve only a single unknown; the undesired effect of coupling between different transport equations does not occur in these cases.

This is different with the conditions (5.27) and (5.28), which together describe the effect of surface recombination. Since recombination clearly involves both carrier types, both electron and hole concentrations resp. Fermi levels occur in the conditions, and the electron and hole transport equations are coupled by these conditions.

The most difficult case is again the non-ideal contact. Although majority and minority carriers are not directly coupled to each other in the current boundary conditions (6.130), (6.131), (6.138), and (6.137), a coupling to the Poisson equation is present in these conditions through the tunneling energy range W_T, which is related to the shape of the band edge and hence to the potential.

9.3 Interface Conditions

According to our respective definition of an interface in Section 2.5, the transport equations have to be solved on both sides of the interface. Interface conditions connect the solutions in the two subregions, taking physical effects at the interface (e.g. thermionic emission or tunneling) into account. The basic types of interface conditions are

$$\mathcal{U}_1 = \mathcal{U}_2 + \mathcal{G} \qquad \text{(ID)}, \tag{9.9}$$

$$\mathcal{F}_1 = \mathcal{F}_2 + \mathcal{G} \qquad \text{(IC)}, \tag{9.10}$$

$$\mathcal{F}_1 = f(\mathcal{U}_1, \mathcal{U}_2) \qquad \text{(IE)}. \tag{9.11}$$

In (9.9), the *unknowns* are connected at the interface, optionally offset by some given parameter. In a certain sense, this is a Dirichlet interface condition (ID). Eq. (9.10) describes flux continuity (IC) at the interface, possibly with some additional surface source $\mathcal{G}$. Eq. (9.11) is of the form of emission currents (IE).

Interface conditions occurred in the previous chapters in (see Table 9.6)

- the general conditions (3.11) and (3.16) for the Poisson equation at arbitrary interfaces (band lineup and continuity of the electric displacement);
- the conditions (7.28), (7.29), (7.30), and (7.31) for the charge and energy continuity equations at the semiconductor heterojunction.

Table 9.6. *Interface conditions*

All interfaces	
$W_{c2} - W_{c1} = \Delta W_c$	(3.11)
$\varepsilon_2 \dfrac{\partial \varphi_2}{\partial x} - \varepsilon_1 \dfrac{\partial \varphi_1}{\partial x} = -\sigma$	(3.16)
Heterojunction	
$J_{2x} = J_{1x}$	(7.28)
$Q_{2x} = Q_{1x} + \dfrac{1}{q}\Delta W_c J_{2x}$	(7.29)
$J_{2x} = q\left[v_{n2}(T_{n2})n_2 - \dfrac{m_2}{m_1} v_{n1}(T_{n1})n_1 \exp\left(-\dfrac{\Delta W_c}{k_B T_{n1}}\right)\right]$	(7.30)
$Q_{2x} = -2\left[k_B T_{n2} v_{n2}(T_{n2})n_2 - \dfrac{m_2}{m_1} k_B T_{n1} v_{n1}(T_{n1})n_1 \exp\left(-\dfrac{\Delta W_c}{k_B T_{n1}}\right)\right]$	(7.31)

Usually, the band lineup ΔW_c in condition (3.11) is taken as a constant parameter, although models exist that take into account a dependence on the local electric field strength. In both cases, however, only the potential (or likewise the band edge) is involved and no coupling to the carrier transport equations is present. In condition (3.16), on the contrary, the interface charge can be modelled as a function of the local Fermi level, which couples the Poisson and continuity equation at the interface.

The interface conditions for hot electron transport at heterojunctions connect the carrier continuity and energy balance equations, since both the carrier concentration and energy density (likewise the quasi-Fermi level and the temperature) or their fluxes occur in the equations (except for the particle conservation equation (7.28), where only the fluxes of the charged particles appear). An additional coupling to the Poisson equation might arise in case of an electric field-dependent model of the band discontinuity.

The simulation domain has to be discretized in each of the subregions adjacent to the interface. We can distinguish two basic configurations for the volume discretization at a material interface. Either the interface passes through the interior of an interface element (type 1) as in Fig. 9.4, or it coincides with the surface of the adjacent elements (type 2); see Fig. 9.5. The type 2 case can be further subdivided into the cases where the node is on the boundary (Fig. 9.5a) or lies in the interior of the element (Fig. 9.5b).

The type 1 interface discretization has been described e.g. in [181, 185]. Implementations of type 2a) are reported, for instance, by Grupen et al. [154] for a semiconductor heterointerface, and Selberherr [48] for a semiconductor-insulator interface. Type 2b) has been used in implementations reported in e.g. [152], [186], and [187].

We have to keep in mind that in principle for each continuity equation

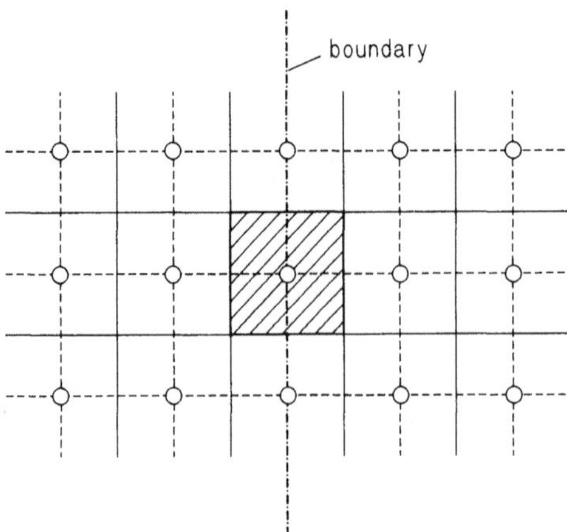

Fig. 9.4 Discretization element located across the interface

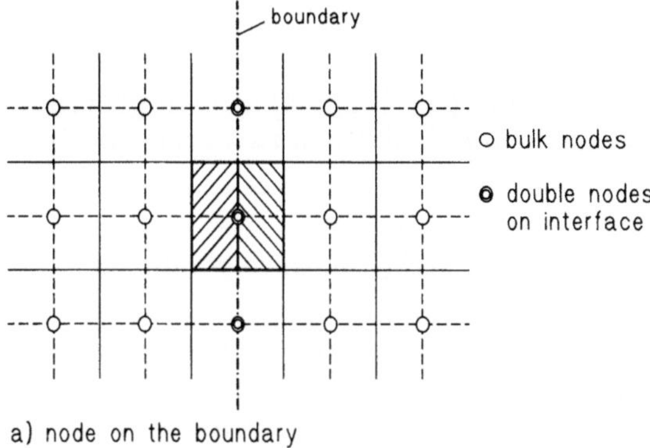

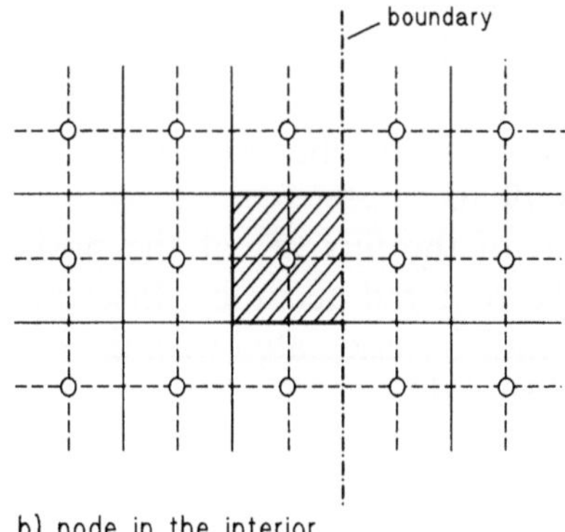

Fig. 9.5 Discretization element touching the interface

there are two interface conditions per interface; each arising from the respective inflow of the two subregions separated by the interface. Hence, in type 1 discretization (Fig. 9.4) the node must hold two variables, one for each subregion. Type 2 discretizations have two elements at the interface; thus each of the two interface conditions must be used to compute the unknowns for one of the two elements.

In the following we shall briefly discuss how the different cases of interface conditions (9.9)–(9.11) can be treated in the various discretization types (1, 2a , 2b). Case IC as a surface version of the continuity equation is always in the game. This interface condition is supplemented by either a condition of case ID or IE to complete the pair.

The following discussion certainly cannot be exhaustive in the sense that other possibilities do not exist. It represents merely a systematic, fairly closed overview of the most commonly used approaches. For instance, equivalent variations of the conditions (9.9)–(9.11) can be easily generated by forming linear combinations, or inserting one condition into either another interface condition or the discrete transport equation.

9.3.1 One Element at the Interface

In the type 1 discretization (see Fig. 9.4), it is in general necessary to consider the node as consisting virtually of two nodes in the same place—one belonging to the first subdomain, the other to the second—so that we have two unknowns for our two equations.

In case ID, we can directly use the interface condition to establish a relation between the values U_1 and U_2 on the two virtual nodes.

In case IC, the discrete conservation equation (9.5) is established for the whole element. Herewith, the fluxes in (9.5) have to be computed using the value U_1 or U_2, according to the respective subdomain where the flux resides. The fluxes at the interface do not occur in this formulation. Since the interface lies in the interior of the element, the surface source $\mathcal{G}$ in (9.10) must be converted into an equivalent volume source. For instance, in case of the Poisson interface condition (3.12) the surface charge σ on the interface passing through the element has the same effect as a space charge $\rho = \sigma/h$, with h according to (9.5).

In case IE, an interpolation of the flux $\mathcal{F}_1$ at the position of the node has to be computed with the required accuracy. Insertion into (9.11) and using the two unknowns U_1 and U_2 associated with the node then gives the respective discrete interface condition.

9.3.2 Two Nodes on the Interface

Now we turn to type 2a. We have two elements that touch each other at the interface. Each of the elements has its node on the boundary. Normally the two nodes share the same position. In this way each node can hold the values of the discontinuous variables in its own region.

Case ID is again no problem; the interface condition (9.9) directly gives the relation between the discrete unknowns.

Case IC can be considered in two ways. Either the interface condition is implemented directly by inserting discrete flux approximations into (9.10), or the conservation equation (9.5) is set up for one of the elements, where the flux across the interface is computed according to the interface condition.

In the former version, accurate approximations for the two fluxes at the nodes are required. This approach has been discussed by Selberherr in [48] for the case of the Poisson interface condition (3.12). He computed the fluxes on the boundary to order $O(h^2)$ by Taylor expansion, and substituted the appearing derivatives of the fluxes with the help of the discrete Poisson equation.

The latter approach, where the conservation equation is solved for the element on one side of the interface, requires an accurate approximation of the flux on the other side, in order to insert it into (9.5). This approximation can be obtained, for instance, with the help of the second interface condition or the conservation equation.

Case IE is straightforward in this discretization. The conservation equation (9.5) for one of the elements is constructed, where the flux at the interface can be directly computed from (9.11) with the variables at the two nodes. This method has been discussed by Grupen et al. in [154] in connection with the computation of the thermionic emission current density at a heterojunction according to Eq. (7.30). In their practical implementation, they reformulated the discrete equation such that the values at the second node are computed online, and no extra storage for that node is necessary. This approach has thus some flavour of the type 1 discretization, too.

9.3.3 Two Elements at the Interface

Finally, we discuss discretization type 2b (cf. Fig. 9.5b). As in the previous subsection, two elements touch each other at the interface. However, now the nodes are no longer on the interface, but lie in the interior of the elements.

For case ID, an extrapolation formula for $\mathcal{U}$ is required in order to compute the unknowns on the interface from the values at the nodes. In this way, the interface condition (9.9) can be posed for the unknowns at the correct position.

Case IC can be treated by invoking the discrete conservation equation for one of the elements, say element 1. The flux F_1 through the interface is substituted with F_2 from the interface condition (9.10). The interface flux F_2, in turn, can be computed from the conservation equation applied to element 2. Hence, in summary the flux balance is computed for the combination of both elements, with the surface source properly accounted for. This treatment is similar as in discretization type 1 (see above).

For case IE, we use again the conservation equation (9.5). In contrast to the previous case, the interface flux is now computed from the variables $\mathcal{U}$ at the interface. Because the nodes are not located on the interface, these values are not available readily, but have to be extrapolated from the nodes as in case ID above.

Appendices

A Transformation of k-Vectors

Many times in the book, an integral over the k-space of one energy band has to be transformed into an integral over the k-space of a band on the other side of an interface, where the k-vectors of both bands are related by energy and transverse momentum conservation. In this Appendix I want to demonstrate the general validity of the relation (4.18)

$$v_{2x}(\mathbf{k}_2)\, d^3\mathbf{k}_2 = v_{1x}(\mathbf{k}_1)\, d^3\mathbf{k}_1, \tag{A.1}$$

which is useful for the frequently occurring transformation of k-space integrals from $\mathbf{k}_1$ to $\mathbf{k}_2$ and vice versa.

The relation of the wave vectors $\mathbf{k}_1$ and $\mathbf{k}_2$ is fixed by energy conservation

$$W_1(\mathbf{k}_1) = \Delta W + W_2(\mathbf{k}_2) \tag{A.2}$$

and conservation of tangential crystal momentum

$$k_{1y} = k_{2y}, \qquad k_{1z} = k_{2z} \tag{A.3}$$

(cf. Eqs. (4.5) and (4.6)). In (4.5), $W_i(\mathbf{k}_i)$ is the (arbitrary) energy dispersion of band i, and ΔW is a possible energy shift between the bands. These three equations determine $\mathbf{k}_2$ as a function of $\mathbf{k}_1$ and vice versa.

The νth component of the group velocity $\mathbf{v}$ in (A.1) is in general given by

$$v_{i,\nu} = \frac{1}{\hbar}\frac{\partial W_i}{\partial k_{i,\nu}}, \tag{A.4}$$

where i denotes the respective band. A relation between v_1 and v_2 can be derived from the $\mathbf{k}_2(\mathbf{k}_1)$-relationship by a partial differentiation of (A.2) with respect to k_{1x}:

$$\frac{\partial W_1}{\partial k_{1x}} = \frac{\partial W_2}{\partial k_{2x}}\frac{\partial k_{2x}}{\partial k_{1x}} + \frac{\partial W_2}{\partial k_{2y}}\underbrace{\frac{\partial k_{2y}}{\partial k_{1x}}}_{=0} + \frac{\partial W_2}{\partial k_{2z}}\underbrace{\frac{\partial k_{2z}}{\partial k_{1x}}}_{=0} \tag{A.5}$$

Since in a partial differentiation with respect to k_{1x} the y- and z-components of $\mathbf{k}_1$ are kept constant, k_{2y} and k_{2z} do not change too because of (A.3), and the indicated terms in (A.5) vanish. Thus, from (A.5) follows

$$v_{1x}(\mathbf{k}_1) = v_{2x}(\mathbf{k}_2)\frac{\partial k_{2x}}{\partial k_{1x}}. \tag{A.6}$$

One easily verifies that $\partial k_{2x}/\partial k_{1x} > 0$ if one remembers that, when solving (A.2) for k_{2x}, the branch of the energy dispersion has to be chosen that implies an equal sign for v_{1x} and v_{2x}. This makes sure that a particle that approaches the interface on one side leaves the interface on the other side. Now we investigate an integral over $\mathbf{k}_2$-space of the form

$$\int d^3\mathbf{k}_2 \, v_{2x}(\mathbf{k}_2)\cdot F(\mathbf{k}_2)$$

with an arbitrary function F of $\mathbf{k}_2$. Transforming the integration variable to $\mathbf{k}_1$, we get

$$\int v_{2x}(\mathbf{k}_2)\cdot F(\mathbf{k}_2)\,d^3\mathbf{k}_2 = \int v_{1x}(\mathbf{k}_1)\left(\frac{\partial k_{2x}}{\partial k_{1x}}\right)^{-1}\cdot F(\mathbf{k}_2(\mathbf{k}_1))\left|\frac{D(\mathbf{k}_2)}{D(\mathbf{k}_1)}\right|d^3\mathbf{k}_1, \tag{A.7}$$

where v_2 has already been substituted on the right-hand side by (A.6). $D(\mathbf{k}_2)/D(\mathbf{k}_1)$ is the functional determinant defined by [75]

$$\frac{D(\mathbf{k}_2)}{D(\mathbf{k}_1)} = \begin{vmatrix} \dfrac{\partial k_{2x}}{\partial k_{1x}} & \dfrac{\partial k_{2x}}{\partial k_{1y}} & \dfrac{\partial k_{2x}}{\partial k_{1z}} \\[2mm] \dfrac{\partial k_{2y}}{\partial k_{1x}} & \dfrac{\partial k_{2y}}{\partial k_{1y}} & \dfrac{\partial k_{2y}}{\partial k_{1z}} \\[2mm] \dfrac{\partial k_{2z}}{\partial k_{1x}} & \dfrac{\partial k_{2z}}{\partial k_{1y}} & \dfrac{\partial k_{2z}}{\partial k_{1z}} \end{vmatrix}. \tag{A.8}$$

Because of (A.3), some of the derivatives become zero or unity:

$$\frac{D(\mathbf{k}_2)}{D(\mathbf{k}_1)} = \begin{vmatrix} \dfrac{\partial k_{2x}}{\partial k_{1x}} & \dfrac{\partial k_{2x}}{\partial k_{1y}} & \dfrac{\partial k_{2x}}{\partial k_{1z}} \\[2mm] 0 & 1 & 0 \\[2mm] 0 & 0 & 1 \end{vmatrix} = \frac{\partial k_{2x}}{\partial k_{1x}}. \tag{A.9}$$

Although k_{2x} depends in general on all components of $\mathbf{k}_1$, the functional determinant surprisingly involves only a single partial derivative. Inserting this result into (A.7), the determinant and the factor of v_{1x} cancel, and we obtain

$$\int d^3\mathbf{k}_2 \, v_{2x}(\mathbf{k}_2)\cdot F(\mathbf{k}_2) = \int d^3\mathbf{k}_1 \, v_{1x}(\mathbf{k}_1)\cdot F(\mathbf{k}_2(\mathbf{k}_1)). \tag{A.10}$$

Since no assumptions on F have been made, (A.10) is valid for arbitrary integrals; (A.1) is thus a short-hand notation for the transformation of integration variables.

For simplicity, the integration domain of the integrals above have not been shown in the formulas. However, it is clear that the integration domain of the $\mathbf{k}_2$-integration in (A.10) must be transformed on the right-hand side in accordance with the transformation (see also the discussion in Section 4.6).

The discussion of this section was restricted so far to the case of transmission. It can be extended to the case of reflection easily. We consider reflection in region 2 from wave vector $\mathbf{k}_2'$ to wave vector $\mathbf{k}_2$. Reflection in region 1 is obtained by a corresponding replacement of the indices.

The energy conservation law now reads (cf. also (4.53))

$$W_2(\mathbf{k}_2) = W_2(\mathbf{k}_2'). \tag{A.11}$$

Since the particle stays inside the same band at reflection, a band edge shift has not to be considered.

The rest of the derivation goes in complete analogy to the derivation above, with $\mathbf{k}_1$ replaced by $\mathbf{k}_2'$. However, in contrast to the discussion of (A.6), the now occurring derivative $\partial k_{2x}/\partial k_{2x}'$ is negative in the present case, because in reflection the branch of the energy dispersion has to be chosen where the velocity turns into the opposite direction. Hence, the derivative cancels against the absolute of the determinant up to but excluding a minus sign. Consequently, the desired result for the transformation of wave vectors at reflection turns out to be

$$v_{2x}(\mathbf{k}_2)\, d^3\mathbf{k}_2 = -v_{2x}(\mathbf{k}_2')\, d^3\mathbf{k}_2'. \tag{A.12}$$

B Conservation of Transverse Momentum

In this Appendix, we want to verify the assumption that the transverse momentum is conserved when an electron crosses a semiconductor interface. As usual, we assume an interface between two materials 1 and 2 with the normal direction along the x-axis.

We start from the effective mass Schrödinger equation for electrons

$$\frac{\hbar^2}{2m_i}\Delta\psi_i + [W - W_c(\mathbf{x})]\psi_i = 0, \tag{B.1}$$

where ψ_i is the wave function in material i, m_i the respective effective mass, W_c the potential energy (i.e. the conduction band edge), and W the energy eigenvalue. The effective mass is assumed to be constant in each of the materials, with a discontinuous transition at the interface. The interface conditions for the Schrödinger equation (see Subsection 4.1.2) are continuity of the wave function

$$\psi_1 = \psi_2 \tag{B.2}$$

and continuity of its derivative

$$\frac{1}{m_1}\frac{\partial \psi_1}{\partial x} = \frac{1}{m_2}\frac{\partial \psi_2}{\partial x}, \tag{B.3}$$

weighted by the respective effective mass in order to conserve the probability current density.

Now we seek the conditions under which

$$\psi_i(\mathbf{x}) = X_i(x)\cdot \exp(jk_y y + jk_z z) \tag{B.4}$$

is a solution of (B.1) and (B.2)–(B.3), i.e. the y- and z-dependence of ψ_i is described by a plane wave with equal tangential wave vector components on both sides of the interface, independent of i. In other words, the tangential wave vector shall not change at the interface.

Inserting (B.4) into (B.1), we find after division by ψ_i and rearranging

$$\frac{X_i''}{X_i} - k_y^2 - k_z^2 = \frac{2}{\hbar^2}m_i[W_c(x) - W]. \tag{B.5}$$

The primes of X_i denote differentiation with respect to x. Inspecting (B.5), we notice that the left-hand side depends solely on x, because k_y and k_z are constants. Hence, in order to be a valid equation, the right-hand side must depend only on x, too. Consequently, we have the *condition that the potential energy must be independent of y and z*, i.e. $W_c = W_c(x)$, as already indicated in (B.5). If this condition is satisfied, the transverse momentum is conserved across the interface.

Eq. (B.5) can be rewritten into the form of a one-dimensional Schrödinger equation

$$\frac{\hbar^2}{2m_i}X_i'' + [W_x - W_{c,i}(x)]X_i = 0 \tag{B.6}$$

for the function $X(x)$, where W_x and $W_{c,i}$ are defined by

$$W_x = W - \frac{\hbar^2}{2m_2}(k_y^2 + k_z^2) \tag{B.7}$$

and

$$W_{c,i}(x) = \begin{cases} W_c(x) - \dfrac{\hbar^2}{2}(k_y^2 + k_z^2)\left(\dfrac{1}{m_2} - \dfrac{1}{m_1}\right), & i = 1, \\[2ex] W_c(x), & i = 2. \end{cases} \tag{B.8}$$

Note that an apparent potential step appears at the interface for nonzero tangential wave vectors if the effective masses are different.

The interface conditions for the one-dimensional equation (B.6) are derived from (B.2)–(B.3) by insertion of (B.4) to give

$$X_1 = X_2 \tag{B.9}$$

and

$$\frac{1}{m_1} X_1' = \frac{1}{m_2} X_2'. \tag{B.10}$$

Hence, the discontinuity of the effective mass appears again in Eq. (B.6) for the x-direction, while it does not affect the consideration of the tangential directions.

In conclusion we find that the transverse crystal momentum is conserved if the potential energy does not change along tangential directions, even if the effective mass is discontinuous at the interface. Under the conditions stated, the tunneling probability can be calculated using the one-dimensional equation (B.6) with W_x as the eigenvalue. The result can be transferred to the 3D case then by substituting W_x with (B.7) in order to obtain a solution of (B.1).

C Calculation of η_T

In this Appendix, the tunneling energy range is calculated from the consideration that the predominant part of the tunneling current is included in the current integral (6.73).

First, we will determine the position v_{xm}^* of the maximum of the integrand. In general, according to (6.56) we will distinguish in cases (a) resp. (b) below whether the semiconductor at the boundary is either in degeneration or non-degeneration. When searching the maximum of the integrand, we have to consider several subsections in the integration range due to the piecewise definitions (6.54) and (6.56) of D and F, respectively. We are going to treat them one after another.

(a) $\psi_B < \eta_T$ (degeneration): Under this condition, we successively check the subranges of tunneling and thermionic emission of (6.54) for the existence of the maximum. Additionally, in subcases (a1) and (a2) we pay attention to the change of $F(v_x^* + \psi_B)$ at $v_x^* = -\psi_B$ according to (6.56).

(a1) $-\eta_T < v_x^* < -\psi_B$ (tunneling): We determine the maximum in this subrange by setting the derivative of the integrand to zero:

$$\frac{d}{dv_x^*}[\exp(-\delta v_x^{*2})(1 - \psi_B - v_x^*)] = 0. \tag{C.1}$$

Solving (C.1) for v_x^* yields

$$v_{xm}^* = -\frac{1}{2}\left[\psi_B - 1 + \sqrt{(\psi_B - 1)^2 + \frac{2}{\delta}}\,\right]. \tag{C.2}$$

To be a valid result, v_{xm}^* of (C.2) has to lie in the interval stated above, which can be recast into the condition

$$v_{xm}^* < -\psi_B \quad \Rightarrow \quad \psi_B < \frac{1}{2\delta}. \tag{C.3}$$

The next subrange to consider is still tunneling, but already in the tail of F.

(a2) $-\psi_B < v_x^* < 0$ (tunneling): In this range, the derivative to evaluate is

$$\frac{d}{dv_x^*}\exp(-\delta v_x^{*2} - \psi_B - v_x^*) = 0, \tag{C.4}$$

and the solution for v_x^* is

$$v_{xm}^* = -\frac{1}{2\delta}. \tag{C.5}$$

The condition for v_{xm}^* of (C.5) for lying in the valid range of subcase (a2) becomes

$$v_{xm}^* > -\psi_B \quad \Rightarrow \quad \psi_B > \frac{1}{2\delta}. \tag{C.6}$$

The last subrange to consider is the thermionic emission range of v_x^*:

(a3) $0 < v_x^*$ (thermionic emission): In this interval, the derivative of DF with respect to v_x^* is always negative; thus a maximum in this range is never possible.

Now we turn to the non-degeneration case.

(b) $\psi_B > \eta_T$ (non-degeneration): In the nondegenerative case, ψ_B lies outside the integration interval. Thus, in contrast to subcases (a1) and (a2) and as in (6.58), a splitting of the tunneling region of (6.54) at $v_x^* = -\psi_B$ does not occur anymore.

(b1) $-\eta_T < v_x^* < 0$ (tunneling): The derivative to evaluate is the same as in subcase (a2)

$$\frac{d}{dv_x^*}\exp(-\delta v_x^{*2} - \psi_B - v_x^*) = 0, \tag{C.7}$$

and also the result is the same:

$$v_{xm}^* = -\frac{1}{2\delta}. \tag{C.8}$$

Since δ is positive according to (6.55), the condition $v_{xm}^* < 0$ is always satisfied.

(b2) $0 < v_x^*$ (therminonic emission): Again as in (a3), the derivative does not change sign in this subrange, and a maximum cannot occur.

Summarily, we have found three different possibilities where the maximum can lie. In degeneration, these are the cases (a1) and (a2), which are distinguished by conditions (C.3) and (C.6). The distinction depends on the parameter δ defined in (6.55), which in turn depends on W_{00}, which by (6.14) again depends on the doping N_D^*. Low doping results in high δ, and by (C.6) and (C.5) the maximum contribution to the integral is at an energy between the top of the barrier and the Fermi energy. At high doping, $1/\delta$ becomes large, and with (C.3) expression (C.2) becomes valid. Now the energy of maximum contribution is near the Fermi energy and still depends only slightly on δ. Thus, according to the statements in the introduction to Chapter 6, case (a2) corresponds to thermionic field-emission, while case (a1) can be associated with field emission. In nondegeneration, only the single case (b1) occurs.

In the following, we compute η_T from (6.75) for the several cases using the maximum values found.

(a) $\psi_B < \eta_T$ (degeneration)

(a1) $\psi_B < \dfrac{1}{2\delta}$ (therm. field-emission): Inserting (6.54) and (6.56) into (6.75) under the current conditions yields

$$\exp(-\delta v_{xm}^{*\,2}) \cdot [1 - \psi_B - v_{xm}^*]$$
$$= \exp(3) \cdot \exp(-\delta \eta_T^2) \cdot [1 - \psi_B + \eta_T]. \tag{C.9}$$

Keeping only the dominant exponential dependence, (C.9) becomes

$$-\delta v_{xm}^{*\,2} = 3 - \delta \eta_T^2, \tag{C.10}$$

and solving for η_T finally gives

$$\eta_T = \sqrt{v_{xm}^{*\,2} + \frac{3}{\delta}} \tag{C.11}$$

with v_{xm}^* defined in (C.2).

(a2) $\psi_B > \dfrac{1}{2\delta}$ (field emission): Under these conditions, (6.75) becomes

$$\exp(-\delta v_{xm}^{*\,2}) \cdot \exp(-v_x^* - \psi_B)$$
$$= \exp(3) \cdot \exp(-\delta \eta_T^2) \cdot [1 - \psi_B + \eta_T]. \tag{C.12}$$

Neglecting again the linear dependence results in

$$-\delta v_{xm}^{*\,2} - v_x^* - \psi_B = 3 - \delta \eta_T^2, \tag{C.13}$$

and solving for η_T yields

$$\eta_T = \sqrt{\frac{\psi_B + 3}{\delta} - \frac{1}{4\delta^2}} \qquad \text{(C.14)}$$

with v_{xm}^* inserted from (C.5).

(b) $\psi_B > \eta_T$ (non-degeneration): In the non-degeneration case, (6.75) reads

$$\exp(-\delta v_{xm}^{*\,2}) \cdot \exp(-v_x^* - \psi_B)$$
$$= \exp(3) \cdot \exp(-\delta \eta_T^{\,2}) \cdot \exp(\eta_T - \psi_B), \qquad \text{(C.15)}$$

which turns into

$$-\delta v_{xm}^{*\,2} - v_x^* - \psi_B = 3 - \delta \eta_T^{\,2} + \eta_T - \psi_B \qquad \text{(C.16)}$$

by taking the logarithm of (C.15). Inserting (C.8) results in a quadratic equation for η_T, having the solution

$$\eta_T = \frac{1}{2\delta} + \sqrt{\frac{3}{\delta}}. \qquad \text{(C.17)}$$

Eqs. (C.11), (C.14), and (C.17) together constitute the result for the apparent barrier lowering W_T. They are combined into a single equation in (6.76).

D Approximation of Surface Mobility

For the evaluation of (8.24) we make an approximation that will enable us to compute the integral analytically. Hence we write

$$I_s = \int_{-\infty}^{\infty} dk_x \left[\alpha + A \exp(-2\sqrt{aB}|k_x|) \right]$$
$$\times \exp(-ak_x^2 + bk_x - b\sqrt{k_x^2 + c}), \qquad \text{(D.1)}$$

where the abbreviations A and B are defined by

$$A = \frac{\alpha(1 - \alpha)\exp(-2b\sqrt{c})}{1 - (1 - \alpha)\exp(-2b\sqrt{c})} \qquad \text{(D.2)}$$

and

$$B = \frac{b}{\sqrt{a}\sqrt{1 + b\sqrt{c}}}\left[1 + \ln\left(\frac{1 - (1 - \alpha)\exp(-2b\sqrt{c} - 1))}{1 - (1 - \alpha)\exp(-2b\sqrt{c})}\right)\right],$$
$$\text{(D.3)}$$

respectively. Further, we approximate the argument of the exponential function in (D.1) by

$$bk_x - b\sqrt{k_x^2 + c} = \begin{cases} b(k_x - \sqrt{c}), & \text{if} \quad k_x < \sqrt{c}, \\ 0, & \text{if} \quad k_x \geq \sqrt{c}. \end{cases} \qquad \text{(D.4)}$$

All these approximations are chosen such that the essential features of the original function are preserved.

The integration range of the integral in (D.1) now can be decomposed according to the distinctions of (D.4) and the sign of k_x; the resulting integrals are all of Gaussian type and can readily be evaluated, giving essentially error functions in terms of ac, $b\sqrt{c}$, and $b/\sqrt{a}$. These expressions can be rewritten with the help of (8.25), (8.30), and (8.31) into more problem specific quantities:

$$ac = u(x), \qquad b\sqrt{c} = \frac{2E_s}{E_x}\sqrt{u(x)}, \qquad \frac{b}{\sqrt{a}} = \frac{2E_s}{E_x}. \tag{D.5}$$

With these substitutions, A and B of (D.2) and (D.3) become

$$A = \frac{\alpha(1-\alpha)\exp\left(-\dfrac{4E_s}{E_x}\sqrt{u(x)}\right)}{1-(1-\alpha)\exp\left(-\dfrac{4E_s}{E_x}\sqrt{u(x)}\right)}, \tag{D.6}$$

$$B = \frac{\dfrac{2E_s}{E_x}}{\sqrt{1+\dfrac{8E_s}{E_x}\sqrt{u(x)}}}\left[1+\ln\left[\frac{1-(1-\alpha)\exp\left(-\dfrac{4E_s}{E_x}\sqrt{u(x)}-1\right)}{1-(1-\alpha)\exp\left(-\dfrac{4E_s}{E_x}\sqrt{u(x)}\right)}\right]\right]. \tag{D.7}$$

Inserting the result of the integrations into (8.28), we finally obtain the degraded mobility near the interface as given in (8.29). The function F occurring in that equation is defined by

$$\begin{aligned}
F(u(x), E_x, \alpha) = A\Bigg\{ &\exp\left(\left[\frac{E_s}{E_x}+B\right]^2 - \frac{2E_s}{E_x}\sqrt{u(x)}\right)\cdot\operatorname{erfc}\left(\frac{E_s}{E_x}+B\right) \\
&+ \exp\left(\left[\frac{E_s}{E_x}-B\right]^2 - \frac{2E_s}{E_x}\sqrt{u(x)}\right) \\
&\times\left[\operatorname{erf}\left(\sqrt{u(x)}-\frac{E_s}{E_x}+B\right)+\operatorname{erf}\left(\frac{E_s}{E_x}-B\right)\right] \\
&+ \exp(B^2)\cdot\operatorname{erfc}(\sqrt{u(x)}+B)\Bigg\}.
\end{aligned} \tag{D.8}$$

Bibliography

[1] E. Wigner, On the quantum correction for thermodynamic equilibrium, *Phys. Rev.*, 40, pp. 749–759, 1932

[2] E. Fick, *Einführung in die Grundlagen der Quantentheorie*, Aula-Verlag, Wiesbaden, 1984

[3] W. Frensley, Simulation of resonant-tunneling heterostructure devices, *J. Vac. Sci. Technol. B*3, pp. 1261–1266, 1985

[4] W. Hänsch, *The Drift Diffusion Equation and Its Applications in MOSFET Modeling* (Computational Microelectronics), Springer, Wien, 1991

[5] F. Buot, K. Jensen, Lattice Weyl-Wigner formulation of exact many-body quantum-transport theory and applications to novel solid-state quantum-based devices, *Phys. Rev. B*42, pp. 9429–9457, 1990

[6] D. Schroeder, Quantenmechanische Gültigkeitsgrenzen der Boltzmann-Gleichung, Seminar at the Technical University of Hamburg-Harburg, Jan. 31, 1991

[7] C. Jacoboni, P. Lugli, *The Monte Carlo Method for Semiconductor Device Simulation* (Computational Microelectronics), Springer, Wien, 1989

[8] N. Ashcroft, N. Mermin, *Solid State Physics*, Holt, Rinehart, and Winston, New York, 1976

[9] A. Marshak, C. van Vliet, Electrical current and carrier density in degenerate materials with nonuniform band structure, *Proc. IEEE*, 72, pp. 148–164, 1984

[10] E. Azoff, Generalized energy-momentum conservation equations in the relaxation time approximation, *Solid-State Electron.*, 30, pp. 913–917, 1987

[11] R. Kishore, Generalized electric current density in materials with position-dependent band structure, *Solid-State Electron.*, 32, pp. 469–473, 1989

[12] M. Lundstrom, *Fundamentals of Carrier Transport* (Modular Series on Solid State Devices, Vol. X), Addison-Wesley, Reading, Mass., 1990

[13] J. Duderstadt, W. Martin, *Transport Theory*. John Wiley & Sons, New York, 1979

[14] C. Cercignani, *The Boltzmann Equation and Its Applications*, Springer, New York, 1988

[15] K. Blotekjaer, Transport equations for electrons in two-valley semiconductors, *IEEE Trans. Electron Devices*, ED-17, pp. 38–47, 1970

[16] C. McAndrew, E. Heasell, K. Singhal, A comprehensive transport model for semiconductor device simulation, *Semicond. Sci. Technol.*, 2, pp. 643–648, 1987

[17] D. Ventura, A. Gnudi, G. Baccarani, F. Odeh, Multidimensional spherical harmonics expansion of Boltzmann equation for transport in semiconductors, *Appl. Math. Lett.*, 5, pp. 85–90, 1992

[18] D. Schroeder, On a consistent formulation of transport equations, distribution function, and mobility, *Semicond. Sci. Technol.*, 5, pp. 435–437, 1990

[19] H. Unger, W. Schultz, G. Weinhausen, *Elektronische Bauelemente und Netzwerke, I*, Vieweg, Braunschweig, 1979

[20] D. Schroeder, Three-dimensional nonequilibrium interface conditions for electron transport at band edge discontinuities, *IEEE Trans. Comp.-Aided Design*, CAD-9, pp. 1136–1140, 1990

[21] A. Marshak, On the nonequilibrium carrier density equations for highly doped devices and heterostructures, *Solid-State Electron.*, 31, pp. 1551–1553, 1988

[22] J. Blakemore, Approximation for Fermi-Dirac integrals, *Solid-State Electron.* 25, pp. 1067–1076, 1982

[23] G. Lautz, *Elektromagnetische Felder*, Teubner, Stuttgart, 1985

[24] C. Dorny, *A Vector Space Approach to Models and Optimization*, Krieger, Melbourne (Florida, USA), 1980

[25] O. Madelung, *Introduction to Solid-State Theory*, Springer, Berlin, 1981

[26] P. Markowich, C. Ringhofer, C. Schmeiser, *Semiconductor Equations*, Springer, Wien, 1990

[27] F. Bechstedt, R. Enderlein, *Semiconductor Surfaces and Interfaces*, Akademie-Verlag, Berlin, 1988

[28] A. Milnes, D. Feucht, *Heterojunctions and Metal-Semiconductor Junctions*, Academic Press, New York, 1972

[29] J. Sune, P. Olivo, B. Ricco, Self-consistent solution of the Poisson and Schrödinger equations in accumulated semiconductor-insulator interfaces, *J. Appl. Phys.*, 70, pp. 337–345, 1991

[30] E. Rhoderick, R. Williams, *Metal-Semiconductor Contacts*, Clarendon Press, Oxford, 1988

[31] J. Tersoff, The theory of heterojunction band lineups, in *Heterojunction Band Discontinuities* (F. Capasso and G. Margaritondo, eds.), pp. 3–57, North-Holland, Amsterdam, 1987

[32] J. Tersoff, Theory of semiconductor heterojunctions: The role of quantum dipoles, *Phys. Rev.*, B30, pp. 4874–4877, 1984

[33] J. Appelbaum, D. Hamann, Surface-induced charge disturbances in filled bands, *Phys. Rev.*, B10, pp. 4973–4979, 1974

[34] C. Kallin, B. Halperin, Surface-induced charge disturbances and piezoelectricity in insulating crystals, *Phys. Rev.*, B29, pp. 2175–2189, 1984

[35] G. Lehner, *Elektromagnetische Felder für Ingenieure und Physiker*, Springer, Berlin, 1990

[36] L. Landau, E. Lifschitz, *Elektrodynamik der Kontinua*. (Lehrbuch der theoretischen Physik, Vol. VIII), Akademie-Verlag, Berlin, 1980

[37] N. Goldsman, J. Frey, Electron energy distribution for calculation of gate leakage current in MOSFETs, *Solid-State Electron.*, 31, pp. 1089–1092, 1988

[38] W. Hänsch, A. Schwerin, A new approach to calculate the substrate current and oxide injection in a metal-oxide-semiconductor field-effect transistor, *J. Appl. Phys.*, 66, pp. 1435–1438, 1989

[39] C. Wang, An improved hot-electron emission model for simulating the gate-current characteristic of MOSFETs, *Solid-State Electron.*, 31, pp. 229–231, 1988

[40] Y. Chen, T. Tang, Numerical simulation of avalanche hot-carrier injection in short-channel MOSFETs, *IEEE Trans. Electron Devices*, ED-35, pp. 2180–2188, 1988

[41] P. Roblin, A. Samman, S. Bibyk, Simulation of hot-electron trapping and aging of nMOSFETs, *IEEE Trans. Electron Devices*, ED-35, pp. 2229–2237, 1988

[42] S. Sze, *Physics of Semiconductor Devices*, John Wiley & Sons, New York, 1981

[43] M. Doghish, F. Ho, A comprehensive analytical model for metal-insulator-semiconductor (MIS) devices, *IEEE Trans. Electron Devices*, ED-39, pp. 2771–2780, 1992

[44] J. Warner, R. M. B. Grung, *Transistors, Fundamentals for the Integrated-Circuit Engineer*, John Wiley & Sons, New York, 1983

[45] E. Spenke, *Elektronische Halbleiter*, Springer, Berlin, 1965

[46] R. Anderson, Experiments of Ge-GeAs heterojunctions, *Solid-State Electron.*, 5, pp. 341–351, 1962

[47] G. Margaritondo, P. Perfetti, The problem of heterojunction band discontinuities, in *Heterojunction Band Discontinuities* (F. Capasso and G. Margaritondo, eds.), pp. 59–114, North-Holland, Amsterdam, 1987

[48] S. Selberherr, *Analysis and Simulation of Semiconductor Devices*, Springer, Wien, 1984

[49] A. Marshak, On the inappropriate use of the intrinsic level as a measure of the electrostatic potential in semiconductor devices, *IEEE Electron Dev. Lett.*, EDL-6, pp. 128–129, 1985

[50] W. Frensley, Wigner-function model of a resonant-tunneling semiconductor device, *Phys. Rev.*, B36, pp. 1570–1580, 1987

[51] Pötz, W., Self-consistent model of transport in quantum well tunneling structures, *J. Appl. Phys.*, 66, pp. 2458–2466, 1989

[52] W. Frensley, Quantum transport calculation of the frequency response of resonant-tunneling heterostructure devices, *Superlatt. & Microstruct.*, 4, pp. 497–501, 1988

[53] K. Jensen, F. Buot, The methodology of simulating particle trajectories through tunneling structures using a Wigner distribution approach, *IEEE Trans. Electron Devices*, ED-38, pp. 2337–2347, 1991

[54] H. Tsuchiya, M. Ogawa, T. Miyoshi, Simulation of quantum transport in quantum devices with spatially varying effective mass, *IEEE Trans. Electron Devices*, ED-38, pp. 1246–1252, 1991

[55] S. Datta, Quantum devices, *Superlatt. & Microstruct.*, 6, pp. 83–93, 1989

[56] C. Lent, D. Kirkner, The quantum transmitting boundary method, *J. Appl. Phys.*, 67, pp. 6353–6359, 1990

[57] H. Liu, Effective-mass hamiltonian for abrupt heterojunctions, *Superlatt. & Microstruct.*, 3, pp. 413–415, 1987

[58] T. Cunningham, R. Barker, L. Chiu, A comparison using a delta-function model of envelope function approximations for quantum wells, *J. Appl. Phys.*, 63, pp. 5393–5397, 1988

[59] K. Hess, G. Iafrate, Hot electrons in semiconductor heterostructures and super-lattices, in *Hot-Electron Transport in Semiconductors* (L. Reggiani, ed.), p. 201, Springer, Berlin, 1984

[60] B. Nag, Boundary conditions for tunneling through potential barriers in non-parabolic semiconductors, *Appl. Phys. Lett.*, 59, pp. 1620–1622, 1991

[61] B. Nag, Boundary conditions for the heterojunction interfaces of nonparabolic semi-conductors, *J. Appl. Phys.*, 70, pp. 4623–4625, 1991

[62] J. Callaway, *Quantum Theory of the Solid State*, Academic Press, New York, 1974

[63] K. Hess, G. Iafrate, Modern aspects of heterojunction transport theory, in *Hetero-junction Band Discontinuities* (F. Capasso, G. Margaritondo, eds.), pp. 451–487, North-Holland, Amsterdam, 1987

[64] R. Mains, G. Haddad, Wigner function modeling of resonant tunneling diodes with high peak-to-valley ratios, *J. Appl. Phys.*, 64, pp. 5041–5044, 1988

[65] R. Feynman, A. Hibbs, *Quantum Mechanics and Path Integrals*, McGraw-Hill, New York, 1965

[66] D. Blochinzew, *Grundlagen der Quantenmechanik*, Harri Deutsch, Frankfurt a. M., 1977

[67] Z. Huang, P. Cutler, T. Feuchtwang, R. Good jr., E. Kazes, H. Nguyen, S. Park, Computer simulation of a wave packet tunneling through a square barrier, *IEEE Trans. Electron Devices*, ED-36, pp. 2665–2670, 1989

[68] H. Maeda, Electron transport across a semiconductor heterojunction, *Jap. J. Appl. Phys.*, 25, pp. 1221–1226, 1986

[69] H. B. Callen, *Thermodynamics and an Introduction to Thermostatistics*, John Wiley & Sons, New York, 1985

[70] M. Mosko, I. Novak, Picosecond real-space electron transfer in GaAs-n-Al$_x$Ga$_{1-x}$As heterostructures with graded barriers: Monte Carlo simulation, *J. Appl. Phys.*, 67, pp. 890–899, 1990

[71] M. Patil, U. Ravaioli, Monte Carlo analysis of real-space transfer in a three-terminal device, *J. Appl. Phys.*, 72, pp. 161–167, 1992

[72] G. Baccarani, M. Wordeman, An investigation of steady-state velocity overshoot in silicon, *Solid State Electron.*, 28, pp. 407–416, 1985

[73] D. Schroeder, The inflow moments method for the description of electron transport at material interfaces, *J. Appl. Phys.*, 72, pp. 964–970, 1992

[74] In memoriam of Farouk Odeh. Personal communication

[75] I. Bronstein, K. Semendjajew, *Taschenbuch der Mathematik*. Harri Deutsch, Frankfurt a. M., 1972

[76] A. Grinberg, Thermionic emission in heterosystems with different effective electronic masses, *Phys. Rev.*, B33, pp. 7256–7258, 1986

[77] S. Fonash, Band structure and photocurrent collection in crystalline and polycrystalline *p-n* heterojunction solar cells, *Solid State Electron.*, 22, pp. 907–910, 1979

[78] C. Crowell, Richardson constant and tunneling effective mass for thermionic and thermionic-field emission in Schottky barrier diodes, *Solid-State Electron.*, 12, pp. 55–59, 1969

[79] C. Crowell, S. Sze, Current transport in metal-semiconductor barriers, *Solid-State Electron.*, 9, pp. 1035–1048, 1966

[80] R. Pierret, C. Sah, An MOS-oriented investigation of effective mobility theory, *Solid-State Electron.*, 11, pp. 279–290, 1968

[81] C. Mailhiot, D. Smith, T. McGill, Transport characteristics of L-point and Γ-point electrons through GaAs-Ga$_{1-x}$Al$_x$As-GaAs(111) double heterojunctions, *J. Vac. Sci. Technol.*, B1, pp. 637–642, 1983

[82] J. Singh, Valley selective tunneling transistor based on valley discontinuities in AlGaAs heterostructures, *Appl. Phys. Lett.*, 55, pp. 2652–2654, 1989

[83] C. Crowell, V. Rideout, Normalized thermionic-field (T-F) emission in metal-semiconductor (Schottky) barriers, *Solid-State Electron.*, 12, pp. 89–105, 1969

[84] F. Beltram, F. Capasso, J. Walker, R. Malik, Memory phenomena in heterojunction structures: Evidence for suppressed thermionic emission, *Appl. Phys. Lett.*, 53, pp. 376–378, 1988

[85] E. Lee, L. Schowalter, Phonon scattering and quantum mechanical reflection at the Schottky barrier, *J. Appl. Phys.*, 70, pp. 2156–2162, 1991

[86] S. Dhariwal, D. Mehrotra, On the recombination and diffusion limited surface recombination velocities, *Solid-State Electron.*, 31, pp. 1355–1361, 1988

[87] B. Sharma, R. Purohit, *Semiconductor Heterojunctions*, Pergamon Press, Oxford, 1974

[88] K. Maeda, I. Umezu, Nonideal J-V characteristics and interface states of an a-Si:H Schottky barrier, *J. Appl. Phys.*, 68, pp. 2858–2867, 1990

[89] S. Sugino, N. Takakura, D. Chen, R. Dutton, Analysis of writing and erasing procedure of Flotox EEPROM using the new charge balance condition (CBC) model, in *Proc. Workshop on Numerical Modeling of Processes and Devices for Integrated Circuits (NUPAD IV), May 31–June 1, 1992, Seattle*, pp. 65–69, IEEE, 1992

[90] W. Eades, R. Swanson, Calculation of surface generation and recombination velocities at the Si-SiO$_2$ interface, *J. Appl. Phys.*, 58, pp. 4267–4276, 1985

[91] C. Riccobene, G. Wachutka, H. Baltes, Two-dimensional numerical analysis of novel magnetotransistor with partially removed substrate, in *1992 International Electron Devices Meeting, Digest of Technical Papers, Dec. 13–16, 1992, San Francisco*, pp. 513–516, 1992

[92] H. Henisch, *Rectifying Semi-Conductor Contacts*, Oxford University Press, London, 1957

[93] J. Blue, C. Wilson, Two-dimensional analysis of semiconductor devices using general-purpose interactive PDE software, *IEEE Trans. Electron Devices*, ED-30, pp. 1056–1070, 1983

[94] M. Reiser, *Zweidimensionale Lösung der instationären Halbleitertransportgleichungen für Feldeffekt-Transistoren*, Diss. ETH Zürich, 1971

[95] M. Reiser, A two-dimensional numerical FET model for DC, AC, and large-signal analysis, *IEEE Trans. Electron Devices*, ED-20, pp. 35–45, 1973

[96] S. Selberherr, A. Schütz, W. Pötzl, Minimos—a two-dimensional MOS transistor analyzer, *IEEE Trans. Electron Devices*, ED-27, pp. 1540–1550, 1980

[97] D. Scharfetter, H. Gummel, Large-signal analysis of a silicon Read diode oscillator, *IEEE Trans. Electron Devices*, ED-16, pp. 64–77, 1969

[98] D. Kennedy, R. O'Brien, Computer-aided two-dimensional analysis of the junction field-effect transistor, *IBM J. Res. Develop.*, 14, pp. 95–116, 1970

[99] H. Henisch, *Semiconductor Contacts*, Clarendon Press, Oxford, 1989

[100] L. Wagner, R. Young, A. Sugarman, A note on the correlation between the Schottky-diode barrier height and the ideality factor as determined from I–V measurements, *IEEE Electron Dev. Lett.*, EDL-4, pp. 320–322, 1983

[101] R. Stratton, Theory of field emission from semiconductors, *Phys. Rev.*, 125, pp. 67–83, 1962

[102] D. Schroeder, An analytical model of non-ideal ohmic and Schottky contacts for device simulation, in *Proc. 4th Int. Conf. on Simulation of Semiconductor Devices and Processes, Sept. 12–14, 1991, Zurich*, (W. Fichtner, D. Aemmer, eds.), Hartung-Gorre, Konstanz, 1991

[103] D. Schroeder, A boundary condition for the Poisson equation at non-ideal metal-semiconductor contacts, in *Proc. 8th Int. Conf. on Numerical Analysis of Semiconductor Devices and Integrated Circuits (NASECODE VIII), May 18–22, 1992, Vienna*, pp. 105–106, Boole Press, Dun Laoghaire, 1992

[104] F. Padovani, R. Stratton, Field and thermionic-field emission in Schottky barriers, *Solid-State Electron.*, 9, pp. 695–707, 1966

[105] J. Crofton, P. Barnes, A comparison of one, two and three band calculations of contact resistance for a GaAs ohmic contact using the Wentzel-Kramers-Brillouin approximation and a numerical solution to the Schrödinger equation, *J. Appl. Phys.*, 69, pp. 7660–7663, 1991

[106] C. Chang, S. Sze, Carrier transport across metal-semiconductor barriers, *Solid-State Electron.*, 13, pp. 727–740, 1970

[107] K. Shenai, Very low resistance nonalloyed ohmic contacts to Sn-doped molecular-beam epitaxial GaAs, *IEEE Trans. Electron Devices*, ED-34, pp. 1642 1649, 1987

[108] S. Flügge, *Rechenmethoden der Quantentheorie*, Springer, Berlin, 1965

[109] A. Schenk, A model for the field and temperature dependence of Shockley-Read-Hall lifetimes in silicon, *Solid-State Electron.*, 35, pp. 1585–1596, 1992

[110] R. Stratton, Volt-current characteristics for tunneling through insulating films, *J. Phys. Chem. Solids*, 23, pp. 1177–1190, 1962

[111] J. Conley, C. Duke, G. Mahan, J. Tiemann, Electron tunneling in metal-semiconductor barriers, *Phys. Rev.*, 150, pp. 466–469, 1966

[112] C. Maziar, M. Lundstrom, Monte Carlo simulation of GaAs Schottky barrier behaviour, *Electronics Letters*, 23, pp. 91–62, 1987

[113] G. Baccarani, Current transport in Schottky-barrier diodes, *J. Appl. Phys.*, 47, pp. 4122–4126, 1976

[114] A. Yu, Electron tunneling and contact resistance of metal-silicon contact barriers, *Solid-State Electron.*, 13, pp. 239–247, 1970

[115] G. Baccarani, Physical models for numerical device simulation, in *Process and Device Modeling* (W. Engl., ed.), pp. 107–158, North-Holland, Amsterdam, 1986

[116] D. Schroeder, T. Ostermann, O. Kalz, Comparison of transport models for the simulation of degenerate semiconductors, Semicond. Sci. Technol., 9, pp. 364–369, 1994

[117] D. Heslinga, T. Klapwijk, Schottky barrier and contact resistance at a niobium/silicon interface, *Appl. Phys. Lett.*, 54, pp. 1048–1050, 1989

[118] F. Guibaly, Surface barrier junctions: Majority- and minority-carrier currents, *J. Appl. Phys.*, 65, pp. 1168–1175, 1989

[119] J. Nylander, F. Masszi, S. Selberherr, S. Berg, Computer simulations of Schottky contacts with a non-constant recombination velocity, *Solid-State Electron.*, 32, pp. 363–367, 1989

[120] S. Sze, *Semiconductor Devices, Physics and Technology*, John Wiley & Sons, New York, 1985

[121] W. Schultz, Zur Theorie der Gleichrichtung am Kontakt Metall-Halbleiter, *Z. Phys.*, 138, pp. 598–612, 1954

[122] C. Crowell, M. Beguwala, Recombination velocity effects on current diffusion and imref in Schottky barriers, *Solid-State Electron.*, 14, pp. 1149–1157, 1971

[123] F. Berz, The Bethe conditions for thermionic emission near an absorbing boundary, *Solid-State Electron.*, 28, pp. 1007–1013, 1985

[124] J. Adams, T. Tang, A revised boundary condition for the numerical analysis of Schottky barrier diodes, *IEEE Electron Dev. Lett.*, EDL-7, pp. 525–527, 1986

[125] J. Adams, A. Jelenski, D. Navon, T. Tang, Numerical analysis of GaAs epitaxial-layer Schottky diodes, *IEEE Trans. Electron Devices*, ED-34, pp. 1963–1970, 1987

[126] R. Darling, High-field, nonlinear electron transport in lightly doped Schottky-barrier diodes, *Solid-State Electron.*, 31, pp. 1031–1047, 1988

[127] R. Stratton, Diffusion of hot and cold electrons in semiconductor barriers, *Phys. Rev.*, 126, pp. 2002–2014, 1962

[128] H. Hjelmgren, Numerical modeling of hot electrons in n-GaAs Schottky-barrier diodes, *IEEE Trans. Electron Devices*, ED-37, pp. 1228–1234, 1990

[129] J. Yoo, H. Lee, Theoretical specific resistance of ohmic contacts to n-GaAs, *J. Appl. Phys.*, 68, pp. 4903–4905, 1990

[130] R. Kupka, W. Anderson, Minimal ohmic contact resistance limits to n-type semiconductors, *J. Appl. Phys.*, 69, pp. 3623–3632, 1991

[131] K. Tomizawa, *Numerical Simulation of Submicron Semiconductor Devices*, Artech House, London, 1993

[132] G. Nanz, A critical study of boundary conditions in device simulation, in *Proc. 4th Int. Conf. on Simulation of Semiconductor Devices and Processes, Sept. 12–14, 1991, Zurich* (W. Fichtner, D. Aemmer, eds.), Hartung-Gorre, Konstanz, 1991

[133] R. Forghieri, A. Guerrieri, P. Ciampolini, A. Gnudi, M. Rudan, G. Baccarani, A new discretization strategy of the semiconductor equations comprising momentum and energy balance, *IEEE Trans. Comp.-Aided Des.*, CAD-7, pp. 231–242, 1988

[134] A. Gnudi, D. Ventura, G. Baccarani, One-dimensional simulation of a bipolar transistor by means of spherical harmonics expansion of the Boltzmann transport equation, in *Proc. 4th Int. Conf. on Simulation of Semiconductor Devices and Processes, Sept. 12–14, 1991, Zurich* (W. Fichtner, D. Aemmer, eds.), Hartung-Gorre, Konstanz, 1991

[135] K. Hennacy, N. Goldsman, A generalized Legendre polynomial/sparse matrix approach for determining the distribution function in non-polar semiconductors, *Solid-State Electron.*, 36, pp. 869–877, 1993

[136] D. Schroeder, D. Ventura, A. Gnudi, G. Baccarani, Boundary conditions for the spherical harmonics expansion of the Boltzmann equation, *Elec. Lett.*, 28, pp. 995–996, 1992

[137] F. Capasso, G. Margaritondo, eds., *Heterojunction Band Discontinuities*, North-Holland, Amsterdam, 1987

[138] K. Hess, Real Space Transfer: Generalized Approach to transport in confined geometries, *Solid-State Electron.*, 31, pp. 319–324, 1988

[139] I. Kizilyalli, K. Hess, Physics of real-space transfer transistors, *J. Appl. Phys.*, 65, pp. 2005–2013, 1989

[140] C. Maziar, M. Klausmeier-Brown, S. Bandyopadhyay, M. Lundstrom, S. Datta, Monte Carlo evaluation of electron transport in heterojunction bipolar transistor base structures, *IEEE Trans. Electron Devices*, ED-33, pp. 881–888, 1986

[141] H. Tian, K. Kim, M. Littlejohn, Novel charge injection transistors with heterojunction source (launcher) and drain (blocker) configurations, *Appl. Phys. Lett.*, 63, pp. 174–176, 1993

[142] S. Luryi, Hot-electron injection and resonant-tunneling heterojunction devices, in *Heterojunction Band Discountinuities* (F. Capasso, G. Margaritondo, eds.), pp. 489–564, North-Holland, Amsterdam, 1987

[143] S. Yngvesson, *Microwave Semiconductor Devices*, Kluwer, Boston, 1991

[144] K. Hess, *Advanced Theory of Semiconductor Devices*, Prentice-Hall, Englewood Cliffs (New Jersey, USA), 1988

[145] S. Luryi, A. Kastalsky, Hot electron injection devices, *Superlatt. & Miconstruct.*, 1, pp. 389–400, 1985

[146] P. Mensz, S. Luryi, J. Bean, C. Buescher, Evidence for a real-space transfer of hot holes in strained GeSi/Si heterostructures, *Appl. Phys. Lett.*, 56, pp. 2663–2665, 1990

[147] H. Schreiber, B. Bosch, Moderne schnelle Silizium-Bipolartransistoren: Heutiger Stand und Tendenzen, in *Heterostruktur-Bauelemente, Vorträge der ITG-Fachtagung 25.–27. April 1990, Schwäbisch Gmünd* (B. Schwaderer, ed.), pp. 49–54, vde-verlag, Berlin, 1990 (ITG-Fachbericht 112)

[148] P. Narozny, D. Köhlhoff, H. Kibbel, E. Kasper, Herstellung und Eigenschaften selbstjustierender Si/SiGe-Heterobipolartransistoren, in *Heterostruktur-Bauelemente, Vorträge der ITG-Fachtagung 25.–27. April 1990, Schwäbisch Gmünd* (B. Schwaderer, ed.), pp. 55–60, vde-verlag, Berlin, 1990 (ITG-Fachbericht 112)

[149] H. Schreiber, B. Bosch, Si/SiGe heterojunction bipolar transistors with current gains up to 5000, in *1989 Internationl Electron Devices Meeting, Digest of Technical Papers, Dec. 3–6, 1989, Washington, D.C.*, pp. 643–646, 1989

[150] A. Das, M. Lundstrom, Numerical study of emitter-base junction design for Al-GaAs/GaAs heterojunction bipolar transistors, *IEEE Trans. Electron Devices*, ED-35, pp. 863–870, 1988

[151] C. Wu, E. Yang, Carrier transport across heterojunction interfaces, *Solid-State Electron.*, 22, pp. 241–248, 1979

[152] S. Mottet, J. Viallet, Thermionic emission in heterojunctions, in *Proc. 3rd Int. Conf. on Simulation of Semiconductor Devices and Processes, Sept. 26–28, 1988, Bologna*, pp. 97–108

[153] K. Horio, H. Yanai, Numerical modeling of heterojunctions including the thermionic emission mechanism at the heterojunction interface, *IEEE Trans. Electron Devices*, ED-37, pp. 1093–1098, 1990

[154] M. Grupen, K. Hess, G. Song, Simulation of transport over heterojunctions, in *Proc. 4th Int. Conf. on Simulation of Semiconductor Devices and Processes, Sept. 12–14, 1991, Zurich* (W. Fichtner, D. Aemmer, eds.), Hartung-Gorre, Konstanz, 1991

[155] M. Mosko, I. Novak, P. Quittner, On the analytical approach to the real space electron transfer in GaAs-AlGaAs heterostructures, *Solid-State Electron.*, 31, pp. 363–366, 1988

[156] M. Mosko, I. Novak, Energy exchange between heterostructure layers by real-space electron transfer, *J. Appl. Phys.*, 66, pp. 2011–2019, 1989

[157] M. Pinto, S. Luryi, Simulation of multiply connected current-voltage characteristics in charge injection transistors, in *1991 International Electron Devices Meeting, Digest of Technical Papers, Dec. 8–11, 1991, Washington, D.C.*, pp. 507–510, 1991

[158] G. Tait, C. Westgate, Electron transport in rectifying semiconductor alloy ramp structures, *IEEE Trans. Electron Devices*, ED-38, pp. 1262–1270, 1991

[159] K. Yang, J. East, G. Haddad, Numerical modeling of abrupt heterojunction using a thermionic-field emission boundary condition, *Solid-State Electron.*, 36, pp. 321–330, 1993

[160] A. Grinberg, M. Shur, R. Fischer, H. Morkoc, An investigation of the effect of graded layers and tunneling on the performance of AlGaAs/GaAs heterojunction bipolar transistors, *IEEE Trans. Electron Devices*, ED-31, pp. 1758–1764, 1984

[161] C. Huang, J. Faricelli, N. Arora, A new technique for measuring MOSFET inversion layer mobility, *IEEE Trans. Electron Devices*, ED-40, pp. 1134–1139, 1993

[162] H. Wong, "Universal" effective mobility of empirical local mobility models for n- and p-channel silicon MOSFETs, *Solid-State Electron.*, 36, pp. 179–188, 1993

[163] S. Selberherr, W. Hänsch, M. Seavey, J. Slotboom, The evolution of the MINIMOS mobility model, *Solid-State Electron.*, 33, pp. 1425–1436, 1990

[164] T. Ando, A. Fowler, F. Stern, Electronic properties of two-dimensional systems, *Rev. Mod. Phys.*, 54, p. 437, 1982

[165] E. Sangiorgi, M. Pinto, A semi-empirical model of surface scattering for Monte Carlo simulation of silicon n-MOSFETs, *IEEE Trans. Electron Devices*, ED-39, pp. 356–361, 1992

[166] M. Tang, K. Evans-Lutterodt, G. Higashi, T. Boone, Roughness of the silicon (001)/ SiO_2 interface, *Appl. Phys. Lett.*, 62, pp. 3144–3146, 1993

[167] D. Ferry, Effects of surface roughness in inversion layer transport, in *1984 International Electron Devices Meeting, Digest of Technical Papers, Dec. 9–12, 1984, San Francisco*, pp. 605–608, 1984

[168] T. Ohmi, K. Kotani, A. Teramoto, M. Miyashita, Dependence of electron channel mobility of Si-SiO_2 microroughness, *IEEE Electron Dev. Lett.*, EDL-12, pp. 652–654, 1991

[169] C. Fiegna, E. Sangiorgi, Modeling of high-energy electrons in MOS devices at the microscopic level, *IEEE Trans. Electron Devices*, ED-40, pp. 619–627, 1993

[170] Y. Ohno, Electron viscosity effects on electron drift velocity in silicon MOS inversion layers, in *1989 International Electron Devices Meeting, Digest of Technical Papers, Dec. 3–6, 1989, Washington, D.C.*, pp. 319–322, 1989

[171] D. Ventura, A. Gnudi, G. Baccarani, Modeling impact ionization in bipolar transistor by means of a direct solution of the BTE, in *Proc. Workshop on Numerical Modeling of Processes and Devices for Integrated Circuits (NUPAD IV), May 31–June 1, 1992, Seattle*, pp. 59–63, IEEE, 1992

[172] J. Schrieffer, Effective carrier mobility in surface-space charge layers, *Phys. Rev.*, 97, pp. 641–646, 1955

[173] R. Greene, D. Frankl, D. Zemel, Surface transport in semiconductors, *Phys. Rev.*, 118, pp. 967–975, 1960

[174] G. Baccarani, A. Mazzone, C. Morandi, The diffuse scattering model of effective mobility in the strongly inverted layer of MOS transistors, *Solid-State Electron.*, 17, pp. 785–789, 1974

[175] K. Fuchs, H. Wills, The conductivity of thin metallic films according to the electron theory of metals, *Proc. Cambridge Phil. Soc.*, 34, pp. 100–108, 1938

[176] M. Schubert, B. Höfflinger, R. Zingg, An analytical model of strongly inverted and accumulated silicon films, *Solid-State Electron.*, 33, pp. 1553–1568, 1990

[177] V. Axelrad, Fourier method modeling of semiconductor devices, *IEEE Trans. Comp.-Aided Design*, CAD-9, pp. 1225–1237, 1990

[178] D. Schroeder, Diskretisierung von Transportgleichungen in 3D-unregelmäßigen Gittern, Seminar at the Technical University of Hamburg-Harburg, Dec. 18, 1989

[179] G. Wachutka, Unified framework for thermal, electrical, magnetic, and optical semiconductor device modeling, *COMPEL*, 10, pp. 311–321, 1991

[180] A. Franz, G. Franz, S. Selberherr, C. Ringhofer, P. Markovich, Finite boxes—A generalization of the finite difference method suitable for semiconductor device simulation, *IEEE Trans. Electron Devices*, ED-30, pp. 1070–1082, 1983

[181] R. Bank, D. Rose, W. Fichtner, Numerical methods for semiconductor device simulation, *IEEE Trans. Electron Devices*, ED-30, pp. 1031–1041, 1983

[182] G. Heiser, *Design and Implementation of a Three-Dimensional, General Purpose Semiconductor Device Simulator*, Hartung-Gorre, Konstanz, 1991

[183] B. McCartin, Discretization of the semiconductor device equations, in *New Problems and New Solutions for Device and Process Modelling* (J. Miller, ed.), pp. 72–82, Boole Press, Dun Laoghaire, 1985

[184] E. Buturla, J. Johnson, S. Furkay, P. Cottrell, A new 3D device simulation formulation, in *Proc. 6th Int. Conf. on Numerical Analysis of Semiconductor Devices and Integrated Circuits (NASECODE VI), Dublin, July 11–14, 1989*, pp. 291–296, Boole Press, Dun Laoghaire, 1989

[185] H. Elschner, D. Bothmann, W. Klix, R. Spallek, R. Stenzel, R. Vanselow, R. Voigt, Methods and results in 2D process and device simulation as well as in 3D device simulation, in *Proc. 6th Int. Conf. on Numerical Analysis of Semiconductor Devices and Integrated Circuits (NASECODE VI), Dublin, July 11–14, 1989*, pp. 148–153, Boole Press, Dun Laoghaire, 1989

[186] S. Mottet, C. Simon, J. Viallet, Flux conservative discretization and adaptive mesh refinement: program CARMES, in *Software Tools for Process, Device and Circuit Modelling* (W. Crans, ed.), pp. 137–142, Boole Press, Dun Laoghaire, 1989

[187] Z. Li, S. McAlister, C. Hurd, Use of Fermi statistics in two-dimensional numerical simulation of heterojunction devices, *Semicond. Sci. Technol.*, 5, pp. 408–413, 1990

Index

N. Arora

MOSFET Models for VLSI Circuit Simulation

Theory and Practice

1993. 270 figures. XXII, 605 pages.
Cloth DM 298,–, öS 2086,–
ISBN 3-211-82395-6
(Computational Microelectronics)

The book covers the MOS transistor models and their parameters required for VLSI simulation of MOS integrated circuits. It gives the first detailed presentation of model parameter determination for MOS models. Various models are developed ranging from simple to more sophisticated models that take into account new physical effects observed in submicron devices used in today's MOS VLSI technology. The assumptions used to arrive at the models are emphasized so that the accuracy of the model in describing the device characteristics are clearly understood. Understanding these models is essential when designing circuits for the state of the art MOS IC's. Threshold voltage being the single most important MOSFET parameter, a full chapter is devoted to the development of the device threshold voltage model. Due to the importance of designing reliable circuits, the device reliability models as applied for circuit simulations are also covered. Since the device parameters vary due to inherent processing variations, how to arrive at worst case design parameters are covered. Presentation of the material is such that even an undergraduate student not well familiar with semiconductor device physics can understand the intricacies of MOSFET modeling. The book serves as a technical source in the area of MOSFET modeling for state of the art MOSFET technology for both praticing device and circuit engineers and engineering students interested in the said area.

Prices are subject to change without notice

Sachsenplatz 4–6, P.O.Box 89, A-1201 Wien · 175 Fifth Avenue, New York, NY 10010, USA
Heidelberger Platz 3, D-14197 Berlin · 37-3, Hongo 3-chome, Bunkyo-ku, Tokyo 113, Japan

W. Joppich, S. Mijalkovic

Multigrid Methods for Process Simulation

1993. 126 figures. XVII, 309 pages.
Cloth DM 198,–, öS 1386,–
ISBN 3-211-82404-9

(Computational Microelectronics)

This book is the first one that combines both research in multigrid methods and a particular application field here - process simulation. It is the declared intention of this book to convince by practically demonstrating the power of the multigrid principle and to establish an example of fruitful interdisciplinary interaction. The introduction to multigrid is therefore strictly directed towards the goal to provide the algorithmical overview one needs to compose optimal multigrid algorithms for evolution problems of process simulation and similar applications. The necessary explanation how and why multigrid works is derived from the roots. So the book preassumes no advanced familiarity with numerical analysis. Additionally a complete strategy to implement different algorithmical components on an adaptive multilevel grid structure is presented. The outlined principle of grid definement and adaptation is based on the control of errors and is reliable as well as general. Last but not least the described strategies are applied to "real life" problems of process simulation. Consequently this book is an important contribution to the interdisciplinary challenge of improving numerical techniques for diffusion problems of process simulation.

Prices are subject to change without notice

Sachsenplatz 4–6, P.O.Box 89, A-1201 Wien · 175 Fifth Avenue, New York, NY 10010, USA
Heidelberger Platz 3, D-14197 Berlin · 37-3, Hongo 3-chome, Bunkyo-ku, Tokyo 113, Japan